THE DOG'S MIND

Also by Bruce Fogle

Interrelations Between People and Pets
Pets and Their People
Games Pets Play
Paws Across London
Know Your Dog
Know Your Cat
The Complete Dog Care Manual
The Complete Dog Training Manual
The Cat's Mind
101 Questions Your Dog Would Ask its Vet
101 Questions Your Cat Would Ask its Vet
First Aid for Dogs
First Aid for Cats
The Secret Life of Dog Owners
The Secret Life of Cat Owners
Encyclopedia of the Dog

THE DOG'S MIND

Bruce Fogle, D.V.M., M.R.C.V.S.

Illustrations by Anne B. Wilson

PELHAM BOOKS

To Jules and her good buddies

PELHAM BOOKS

Published by the Penguin Group
Penguin Books Ltd, 80 Strand, London WC2R 0RL, England
Penguin Putnam Inc., 375 Hudson Street, New York, New York 10014, USA
Penguin Books Australia Ltd, 250 Camberwell Road, Camberwell, Victoria 3124, Australia
Penguin Books Canada Ltd, 10 Alcorn Avenue, Toronto, Ontario, Canada M4V 3B2
Penguin Books India (P) Ltd, 11 Community Centre, Panchsheel Park, New Delhi – 110 017, India
Penguin Books (NZ) Ltd, Cnr Rosedale and Airborne Roads, Albany, Auckland, New Zealand
Penguin Books (South Africa) (Pty) Ltd, 24 Sturdee Avenue, Rosebank 2196, South Africa

Penguin Books Ltd, Registered Offices: 80 Strand, London WC2R 0RL, England

www.penguin.com

First published in hardback 1990
First published in paperback June 1992
22

Copyright © Dr Bruce Fogle, 1990

Typeset in Monophoto 10½/12½ pt Photina
Printed in England by Clays Ltd, St Ives plc

A CIP catalogue record for this book is available from the British Library

The moral right of the author has been asserted.

Library of Congress Catalog Card Number: 91-33711

ISBN 0 7207 1964 X

Illustrations on pages 8, 15 and 48 by Kenneth Smith

Contents

Acknowledgements

A personal thank you to my head nurse and people-minder, Jenny Berry. I had worried that an intermittent absence from my veterinary clinic while researching and then writing this book might have had a deleterious effect on the practice. I should have known better. Thanks too, to Roger Abrantes in Denmark for sending me a copy of his book Hundesprog and for his permission for Anne Wilson to use his illustrations as inspiration for some of the marvellous sketches that follow.

Introduction

There are millions of species of life that exist on earth today but of all of these the dog is almost certainly the animal that is closest to our hearts. Dogs are, of course, a fact of life in certain cultures, primarily European, American and Japanese. In these cultures we have lived with them for thousands or even tens of thousands of years. We have allowed them to share our dens, our food, our companionship. Of all the hundreds of millions of species that have ever existed on this earth, surely the dog has become the one we should understand the best.

Yet we have a problem for there is something about dogs that makes us irrational. Ask a cat owner what he enjoys about his cat and more likely than not he'll tell you that he enjoys observing the natural quality of its behaviour. Cat owners – cat LOVERS – are observers and draw a clear line between themselves and their pets. In fact we do so with all species of animals yet, for some reason, we blur it with dogs. 'She's my best friend', I'm told almost daily in veterinary practice. 'I love her as much as my children', pet owners will confide. 'He's part of the family', I universally hear. And let me lay MY cards on the table now. I'm also one of the 89 per cent who talk to their dogs and think of them as members of the family, but in doing so, in creating little furry people out of our canine companions, we lose the ability to understand them as they really are. We think of their behaviour in human terms.

Just as often as I'm told of the love and devotion my clients have for their pets, so too do I hear of the fidelity, love and everlasting affection that my patients have for their owners. And it's not just average pet owners who think this way either. 'Every dog that ever followed its master (gives) an immeasureable sum of love and fidelity,' wrote the Nobel prize winning ethologist Professor Konrad Lorenz in his book *Man and his Dog*.

Konrad Lorenz, in attributing the feeling of 'love' to dogs, was talking about how they feel – about what goes on in a dog's mind, but exactly what is the mind? Can a dog really think? Do dogs have a culture? What is canine intelligence and how should it be defined? These are quite basic questions that I feel I should try to explain now so that you will know what terms of reference I'm using.

First of all, I should explain that I have quite intentionally not called this book, 'The Dog's Brain' or 'The Dog's Behaviour'. It's called 'The Dog's Mind' for several reasons. The brain is simply one of the body's organs – a brilliant and poorly understood organ. In each brain there are billions of cells all with specific functions. In my dog's brain there are more cells than there have ever been dogs! These cells produce their own drugs such as the endorphins, the body's natural pain killers, and these drugs, in turn, affect the 'mind'. The word 'brainy' also denotes intelligence and we naturally assume that we are superior to all other animals because we're the 'brainiest' animals around. That's one of our little conceits. Our brains are certainly different to other animals and have allowed us to become the dominant species on earth, but to argue that our brains are superior is the same as arguing that a cow's intestines are superior to ours because they can digest cellulose fibre.

The philosopher Thomas Nagel once wrote an article called, *What is it like to be a bat?* in which he discussed the philosophical problems of imagining what it is like to be what you are not.

He depicted the bat's ability to echo locate, something that is so alien to our abilities that it's almost impossible to understand, and described how difficult it is for us to imagine that another animal is actually BETTER than we are at something. Nagel's argument applies to the dog too. The dog's brain has an ability to interpret scents and smells that is infinitely different to ours. You could say 'superior' but that is simply a value judgement. They simply can't be compared.

I've avoided the word 'behaviour' in the title for similar reasons although I will in fact use it a great deal in what I describe in the forthcoming chapters. 'Behaviour' brings to mind rats in Skinner boxes, conditioned to press buttons to get food rewards. When the Russian Pavlov conducted his original experiments on dogs, the experiments where, for example, he discovered that a dog could be made to salivate at the sound of a ringing bell if that dog had been trained to associate the sound of the bell with food, he too had a problem in describing what was going on in what part of the dog's body. Other languages don't necessarily have words that are synonymous with the English word 'mind'. In French, 'tête', 'intelligence' or 'esprit' only come close. Pavlov initially used the Russian word 'ym' but later changed this to a phrase which in translation means 'higher nervous activity'.

A further problem with using the word 'mind' is its association in Christian culture with the soul. A few years ago, I conducted a survey of

British veterinarians concerning their attitude to pet death. One out of five practising veterinarians believed that a dog has a soul and an afterlife. (Two out of five believed that humans have souls and an afterlife.) When the same survey was applied a year later to practising veterinarians in Japan (where the Buddhist and Shintoist traditions allow for an afterlife for all living things), every single veterinarian surveyed believed that dogs have souls and an afterlife!

I've avoided all of this and used the word 'mind' intentionally. To me, the dog's mind is a function of its brain, of evolution, of genetics, of the senses, of hormones and of learning and I will discuss each influence in its own chapter. Because learned behaviour is what we have most control over I will devote several chapters to this influence on the dog's mind.

The mind can feel elation or depression, anger, sadness, thirst or hunger, pain or exhilaration. Dogs are sentient beings, aware of their own personalities. They have minds as much as we have and that's why I've used the word, even if its use waves like a red flag at some people. This isn't a dictionary definition but is rather my own definition and is based on the difference between objectivity and subjectivity which I feel also needs an explanation.

In 1953, on Koshima Island in Japan, a monkey named Imo discovered that she could rub the mud off the sweet potatoes she had been given if she washed them underwater. This behaviour, potato washing, soon spread to her playmates, her mother and aunts and later to her own infants who also copied her. By the 1960s, over half of the entire troop of Koshima monkeys washed their potatoes before eating them.

Imo's activities were, of course, observed by scientists who were interested in monkey behaviour but when they published their observations, they were severely criticized by Western animal behaviourists for being too anthropomorphic, for attributing too much human behaviour to their monkeys.

Classically, the objective scientist has never interpreted what an animal perceives or thinks. In fact, animal behaviourists in the West almost universally adhered to this non judgmental approach. They were the Skinner box brigade who gave their animals code names and numbers and who tried to be as rational as possible.

Japanese scientists, and under the influence of Nikko Tinbergen and Konrad Lorenz, other scientists in Europe, continued their more subjective study of animals by getting to know them. The Japanese described their approach as 'kyokan', a term that is difficult to translate into English but which suggests an empathy and an understanding with the animal that is being studied. Their perseverance was of course correct and by the 1970s, the Japanese pattern of long-term studies of individual animals, 'kyokan', became the norm in the western world too.

My approach throughout the coming chapters will be both objective and subjective. I'll describe objective research into the dog's mind but I will also liberally describe subjective and anecdotal observations. By definition, that means that I will discuss, to some extent, how a dog thinks. 'Thinking' brings into question whether dogs have a culture, what is cognition and what is intelligence? Aristotle said that animals can learn and can remember but cannot think. Washoe, the chimpanzee that was taught American sign language, has forever banished the thought that animals are incapable of thinking. Washoe taught us that chimpanzees think a lot – so much so that they can actually lie and be deceitful. But it's not just primates that think.

Ask any shepherd and he will tell you that his sheepdog thinks about what he is going to do. In fact, ask any veterinarian and he will say the same. Dogs dream. During their sleep they go through periods of rapid eye movements (REMs) that on the electroencephalograph are the same as those we experience when we are dreaming. Dogs grieve. This is something that objectively can never be determined but subjectively is apparent to all keen observers of canine activity.

The questions of culture and intelligence, however, are more difficult to define. John Tyler Bonner, in his book, *The Evolution of Culture in Animals*, defines culture as, 'the transfer of information by behavioural means'. Using his definition then, Imo the monkey was creating a new culture when her troop learned to copy her and wash their sweet potatoes underwater.

All is not as clear with dogs. Certainly there are dramatic 'cultural' differences between breeds of dogs. Some are more dexterous than others. Some are fiercer than others. Some are louder and more gregarious than others, but in most instances these 'cultural' differences are as a result of genetic changes, genetic changes which occur under our influence. There is no question that one dog can observe another dog doing something and that it can copy it, but most differences in 'culture' are, in the dog, genetic rather than through learning. That means that to a great extent canine 'culture' is under our control. When left to nature, when dogs are allowed to revert to basic behaviour, almost invariably the result is a dingo-like animal, a dog with a basic culture. We put an extra veneer of culture on the dog's mind through selective breeding and imprinting certain characteristics.

Calculating intelligence in dogs is much more complicated than in other species simply because of the great morphological range within the species. Intelligence can be called the ability to use or call upon past experience in order to adapt to a new experience or situation. But intelligence is not simply an efficiency in adaptive behaviour. All living animals have evolved to be good at this. It doesn't mean smart, clever or academic either. Nor does it mean dexterous. Dogs can be taught to operate handles and bars but little more. But what is it that we really measure when we talk about canine intelligence?

Intelligence tests are always tailored to a species but how can you compare a Chihuahua with a German shepherd. Different breeds have different stamina, sensory perceptions, size, agility and, yes, emotions. We breed dogs for different functions and this makes any comparison of intelligence almost ridiculous. Canine TRAINABILITY and canine OBEDIENCE, on the other hand, are easily measured, which is why there isn't a police force in the world with brigades of Scottish terriers on guard.

Different breeds have different stamina, sensory perception, size, agility and emotions.

However, ALL dogs have magnificent cognitive powers. Cognition is defined as the brain processes through which a dog acquires information about his environment. The dog's mind is eminently good at cognition. They are amazingly perceptive to nuance and observe the most imperceptible changes in us. That's how they know you are going on holiday days before the bags come out, or how they know you'll be taking them to the vet, before you even reach for the car keys. The dog's mind works this way by watching our body posture and he will express what's on his mind through the position of his ears and eyes, how he holds his tail or head and how he moves (see Chapter Five).

What finally makes dogs so difficult to understand is that in many ways they are actually very similar to us. Pack dogs live in groups of mutually interrelated individuals who are each independent but who work as a unit for a common good. They hunt in packs. They enjoy body contact and relish the companionship initially given by their mothers and then latterly by their peers and then others. Social activity is important for them and playful mock aggression, sex and exploration are all a part of their behaviour and ours. The lucid and articulate Stephen Jay Gould believes that there is an analogy between childhood wonder and adult creativity. He says that we humans are neotenized apes – that is, we are apes that have maintained juvenile characteristics into adulthood, both physical and mental characteristics, and that it is these juvenile characteristics that have been responsible for our success as a species. In selectively breeding dogs as we have, there is no doubt that we have neotenized them too. Both morphologically and mentally, dogs have been bred to retain the juvenile characteristics of play, exploration and subservience to the leader. Now, of course, the root stock of the dog, the wolf, already has in its developmental repertoire a whole range of neotenized behaviours. Gould argues that neoteny can be a natural occurrence in animals other than the human, but isn't it curious that of all the species we chose to blur the barrier with, the one we have chosen is the one that, like ourselves, perpetuates juvenile behaviour throughout adult life? That is probably an answer to the question, 'Why dogs?'. What this also means is that to understand the dog's mind we must first briefly look at where the dog has come from, the wolf.

Dogs are wolves, although sometimes they look like they are in sheep's clothing. We have altered their morphology, creating dogs that look like big sheep (the Maremma or Pyrenean Mountain Dog) so that they can live with the flock but protect them from danger. The dog might look like a sheep. It might actually THINK it's a sheep, but the entire basis for its behaviour is what it has inherited from the wolf.

Wolves are opportunists. They evolved to fill the niche of attacker and scavenger and although the true wolf, Canis lupus, is a mammal of the northern hemisphere, other wolf-like animals evolved elsewhere in the world to fill the same niche. Certainly the oddest was the thylacine, the Australian pouched wolf, a marsupial – related to the kangaroo but wolf-like in body and mind. The last one died in a zoo in the 1930s. (Australian dingos are true dogs brought to Australia by the Aborigines around 10,000 years ago.)

Although it has been postulated by Michael Fox and by others that the domestic dog, Canis familiaris, is the descendant of a now extinct European dingo-like wild dog, no fossil of this animal has ever been found and all genetic, behavioural and anatomic evidence firmly supports the belief that our dogs are descended from a small subspecies of wolf that lived in the Near East in the southern part of the wolf's Old World range. The wolf, Canis lupus, is one of the most pleomorphic of all mammals alive today. They have great physical elasticity with the largest breeds of wolf in the north and the smallest in the southern part of their range. Canadian Arctic Tundra wolves can weigh up to perhaps 200 pounds. One weighing 175 pounds was once shot by a forest ranger near Jasper in the Canadian Rockies, not far from the site of the 1988 Winter Olympics. The Arabian desert wolf, on the other hand, weighs only 45 pounds.

Although there were probably over thirty breeds of wolf at the turn of the century at least seven of these have become extinct since then. The massive white Newfoundland wolf, over six feet long, died out early in the century. So too did the small Japanese wolf, an animal only 35 inches long and 14 inches at the shoulder. The specimen in the British Museum has a 12 inch long dog-like tail.

The Texas red wolf, a 40–60 pound breed, 28 inches at the shoulder, has become extinct in the last decade. The German shepherd-size Texas grey wolf and the much larger cinnamon coloured Cascade Mountain wolf have also gone.

The extinction of these breeds is a pity because aside from their different morphological characteristics, each breed has behavioural differences too. The small Asian wolf, Canis lupus pallipes, doesn't howl. The Chinese wolf, Canis lupus laniger, hunts alone. From old stories and fables it appears that the European wolf, Canis lupus lupus, hunted in large packs and was content to live within close proximity to man.

The domestic dog, Canis familiaris, has the same number of chromosomes as the wolf, seventy eight, and this is different from other related canids. Jackals, for example, have seventy-four chromosomes. Red foxes have thirty-eight. Dogs were probably first domesticated from one breed of wolf, the small Near Eastern breed. The most tractable were bred from, the least tractable eaten, and then over the ages, local admixtures of genes from other breeds of wolf augmented the diversity of the dog leading to the great genetic variation of today. This change is still going on naturally amongst wild canines. Dingos in Australia are breeding with domesticated dogs, irreversibly altering for the future how dingos think and behave. The same is happening in the State of Maine in the United States. There the indigenous wolf no longer exists. In its place is a composite wild animal, part wolf, part coyote, part domestic dog.

Human intervention in breeding has meant that the dog has smaller teeth and shorter jaws than the wolf, probably indicating a decreased natural selection for ability to catch and kill. (Little Red Riding Hood was certainly right about the wolf. What big teeth they really do have!) Early dogs undoubtably helped in the hunt but just as undoubtably they benefited from humans feeding them too. Interestingly the frontal sinuses on the dog's head are more inflated than those of the wolf – the snout slopes down more. There doesn't seem to be any reason for this other than perhaps an ages' old subconscious human preference for dogs that looked intelligent. This is a curious fact today where breeds with prominent frontal bones are said to look more intelligent than others. Indeed, it has been interesting to see that amongst golden retrievers, the breed that I have, those with more prominent frontal sinuses almost invariably win at dog shows although this is not the case in field trials where looks are not as important.

Breeds of dogs evolved from the adaptive genetic pool already present in the wolf and have perpetuated the high degree of social organization and communication that stabilizes communal relationships in the wolf pack. Because the wolf pack works as a group of individuals all unique but all working towards the same goal, pack society always allowed different personalities to develop. One wolf might hunt best. Another might be the best strategist.

Adolph Murie, after years of studying wolves in the wild, wrote that his strongest impression was of the wolves' friendliness to each other. In his books Murie describes how adult wolves play tag, how they romp with a rocking horse gait as they play with pups. Murie mirrors Stephen Gould's contention that some species of animal are naturally neotenized and describes how adult timber wolves will suddenly jump out of hiding places and scare each other for no apparent reason other than play, how they bring things to each other, especially food, and how they will prance and parade around with sticks in their mouths. This is the genetic, sensory, morphological, hormonal and evolutionary root stock of the dog and to understand the dog's mind it is absolutely essential to understand that he has come from here. But how?

Earlier this century Russian fox breeders discovered, almost unwittingly, that it was possible to tame silver foxes within twelve generations of controlled breeding, simply by selecting for a stable temperament and tractability. (We humans have been breeding dogs for at least 1000 generations.) John Peters at the University of Arkansas did the exact opposite with pointer dogs. He produced UNSTABLE pointers within a generation. Starting with 'Allegheny Sue', a promising show bitch who produced 'spooky' pups, Peters produced a line of nervous, neurotic and unstable bird dogs. Allegheny Sue's pups were more timid, fearful and panicky than other pointer pups. If, for example, a red fire extinguisher was stood beside their food bowls they wouldn't eat. The pups were this way for at least two reasons. The first was their genetic make-up. These pups had inborn emotional and cognitive problems inherited from their mother and I will discuss this in the next chapter. The second reason was that Allegheny Sue was simply a rotten mother and imprinted unwanted characteristics onto her pups during their first weeks of life. I will discuss maternal imprinting in Chapter Six.

The fact, however, is that it takes very little time to even unwittingly select for temperament in the canine. Twenty thousand or so years ago, when dogs were first domesticated, those with a tendency to bark were selected as guards, the fast and silent ones as hunters. (With our knowledge today it is easier and faster to select for temperament. The Pyrenean Mountain Dog is a classic example. Bred for hundreds of years as a guard and protector of mountain sheep, this powerful dog has been, through selective breeding over the last twenty years, 'modified' to become acceptable as a companion animal with a more benign and giving behaviour.) It was in this way that different canine cultures emerged – hunters – herders – guarders – and then, more recently in the last 200 years, breeds that would flush, point, corner, retrieve or sit quietly on satin cushions.

Different climates, geography, predators and parasites also influenced the production of different breeds or races of dogs as did our own intervention in their reproductivity. Certain aspects of wolf behaviour became modified. Dogs are more fertile than wolves. A she wolf has one reproductive cycle a year while most dog breeds (except the Basenji) have two. Dogs also reach puberty earlier than wolves do and have larger litters, almost certainly, in part, because we help raise them. Most importantly, the socialization period of dog pups is longer than that of wolf pups making it easier for us to train them.

In almost all breeds of dogs we have taken basic wolf characteristics and accentuated them to such a degree that in most instances the dog has become better than the wolf at a specific trait. Bloodhounds can follow scent better than wolves can. Salukis, Borzois and Greyhounds are faster. Terriers are more aggressive, Border collies are better at chasing. German shepherds are better at guarding and attacking. We have improved some traits by sacrificing others and in doing so have altered how the dog thinks. We have altered the dog's mind. In some cases we have gone almost full circle and have produced a breed of dog that looks surprisingly similar to the wolf but is still a manageable and reproductive dog. The Malamute and German shepherd are examples.

Some breeds might physically look similar to wolves but their minds have been significantly modified.

Civilization has been fantastically permissive of the variations in temperament and mental characteristics in dogs compared to 'the wild'. Civilization has allowed for these genetic variations and configurations for thousands of years. There are images of Saluki-like dogs on 7500-year-old Egyptian pottery and Mastiff-like dogs on 4000-year-old Babylonian sculptures. (In Israel, a 12,000-year-old skeleton of a man buried with his arm around a four to five month old dog pup suggests that there has been a bond of affection for at least that length of time.) Xenophon, the Greek historian described 2500 years ago, dogs that chased prey then stood still looking at the animal quivering with excitement. He might just as well have been describing the behaviour of Allegheny Sue, although in his time these 'pointers' were originally exploited by falconers. Dogs that look like Maltese terriers have also existed for at least 2500 years and canines that look like the Pekinese have been around for at least 2000 years. Inuit sled dogs – Malamutes – have almost certainly retained their characteristics for thousands of years.

In our homes too, dogs have existed for centuries. Chaucer, in *The Canterbury Tales* has the Wife of Bath metaphorically describe a woman's over-affection for a man as, 'For as a spaynel she will on him lepe.' Spaniels have remained true to form for a thousand years! Richard Blome, in 1686, in his book *The Gentleman's Recreation* said that, 'Spaniels by Nature are very loveing, surpassing all other Creatures, for in Heat and Cold, Wet and Dry, Day and Night, they will not forsake their

Master.' Konrad Lorenz was obviously not the first writer to infer love into the dog's mind.

In the following chapters, I will describe how the dog's mind is a result of instinct, genetics, evolution and selective breeding, how hormones influence the mind and how maternal and peer imprinting and human intervention alter the ways of the dog. I will also discuss how we can practically use this information and, finally, discuss how the dog's mind changes with illness and more importantly, with old age. The mind is a nebulous thing. Mark Twain once said he had such a prodigious quantity of mind that it often took him as much as a week to make it up. However we approach the dog's mind we must always be aware that we can never really know what goes on in it. We can only hypothesize. This restriction on our understanding applies to all living animals, but with an expanded appreciation of what influences and controls behaviour, we can perhaps hypothesize from both a scientific and sympathetic position. Before I continue let me give you one final example of the dilemma we have.

A cow elephant died suddenly one Saturday morning at a Safari Park and a veterinary pathologist, an expert in wildlife pathology, was called in to determine the cause of death.

The pathologist knew that elephants, especially those that wintered in Florida, could suffer from a virus infection that damaged the heart muscle so suddenly and severely that it could cause unexpected heart failure and instant death. He also knew that there are many other causes of sudden death so he proceeded with a full post mortem.

A mature cow elephant weighs over 3000 kilograms (6600 pounds). Her head alone can weigh over 800 kilograms (1760 pounds) and because there was no practical way of taking the body to the veterinary college, he carried out the post mortem in the shed where she had died.

Soon he had a problem. Pieces of elephant lay around him – too large and too heavy to be moved by the people that were helping him. He discussed this problem with the owner of the Safari Park who called in a bull elephant, the cow's consort, to help move the pieces.

The bull elephant was brought in and told to pick up a leg and carry it to the corner. The pathologist saw that the bull was agitated, but it did as it was directed. The elephant was then asked to roll the cow's head out of the pathologist's way. The bull elephant got more agitated. He swayed from side to side and shifted his weight from leg to leg. The pathologist said that the elephant looked anxious but once more it did as it was told but this time he flailed the air with his trunk and trumpeted loudly when he had finished.

It was apparent to everyone in the shed that the bull elephant was deeply distressed and the trainer opened the shed door to let him out. As soon as the door opened he ran at full speed as far as he could to the farthest ground by the perimeter fence where he pressed his head into the ground and trumpeted loud and long. He stayed there and didn't move until his trainer reached him, spoke to him and stroked him.

I've told this story because it raises the question of how we can EVER know what really goes on in an animal's mind. The objective scientist will only describe what happened. The subjective scientist, and subjectivity, which has blown hot and cold in the behavioural sciences is fortunately blowing hot again, will be more open, will employ 'kyokan' in his understanding of an animal's mind. This applies to our pet dogs too. We can never get into their minds to truly understand how they think but we can surmise in as scientific but also in as empathetic a way as possible. William James summed up our problem. 'Marvellous as may be the power of my dog to understand my moods, deathless as is his affection and fidelity, his mental state is as unsolved a mystery to me as it was to my remote ancestor.' What follows will hopefully unravel the mystery a little.

PART ONE

The Anatomy and Physiology of the Dog's Mind

Chapter One

The Genetics of the Mind

As I'm writing this my two retrievers are in the back garden. One of them has the other pinned to the ground by her neck and gives all the appearances of killing her. Both are making deep growling noises and have been 'attacking' each other now for about ten minutes. Curiously though, it is the dominant older dog that is pinned to the ground by the less dominant younger one. This is in fact something that the older dog seems to intentionally allow the younger one to do.

While they are wrestling, an apple falls from the tree landing on the younger's tail. It sufficiently startles my younger dog to make her cry out, run to the kitchen door and look through, pleadingly, at me. Her pleading look is very effective simply because, with her light hair colour and deep black pigmentation around her brown eyes, she looks like she is wearing more mascara than is proper. The older dog has in the meantime picked the apple up in her mouth and she too is now at the door, apple in mouth, waiting for me to turn the handle and let them in. Are these really wolves in disguise?

We humans have been interfering in dog breeding for thousands of years. Put in another way, for thousands of canine generations, and a canine generation is two or three years compared to the twenty or so for a human generation, we have altered natural genetic selection in dogs. One of the results on this interference with genetic selection is the apple bearing dog standing outside my kitchen door right now.

Genetics, left to nature, selectively produces dogs that can find their own food, have the greatest potential to escape from danger and that will breed with the largest number of females. In other words a dingo-like dog.

Our influence on genetic selection has been to produce dogs that are more manageable, more docile, easier to handle, more fertile and that have a reduced fight or flight response. Through genetics, we have unintentionally selectively bred dogs that perpetuate their juvenile behaviour into adulthood but exactly how has this happened? How have we so affected the dog's mind?

Curiously, few animals have as closely accompanied us as we have undergone the cultural and environmental changes of the last 10,000 years. As human culture evolved from the cave to the campsite, to the farmstead, to the village, to the town and finally now to the urban conglomeration, only the dog and perhaps the cat have moved with us because we have actually wanted them to. The other animals that have accompanied us, rats, mice and birds, have mostly been unwanted opportunists. Technically speaking, the dog has been marvellously successful at ADAPTIVE RADIATION, at populating these new environments with its descendants. It has been so successful because of the brilliant genetic databank that the wolf built up in its cells, a databank that allowed the dog to alter both its mental and physical characteristics as it accompanied man around the world. As the dog moved through cultures, from the forest to the sofa, he had the genetic plasticity to cope well with each new situation.

The dog has been a marvellously successful opportunist.

Genetically speaking, the ideal condition for rapid changes occurs when a species is divided into individual isolated subpopulations which only occasionally meet. Only breeding at these infrequent meetings will allow genetic exchanges between them. This, of course, is exactly what happened and indeed is still happening with dogs. Originally, dogs were isolated as subpopulations, just as we were, by geography. Today, subpopulations of dogs continue to exist through our creation of hundreds of individual breeds.

The notable consequence of this isolation was that in different parts of Asia and Europe, dogs with differing characteristics emerged, gazehounds, war dogs, shepherd dogs, guard dogs. Each type of dog behaved like a dog but excelled in its own particular way. (In North America, dogs evolved too. Indian dogs were as different looking from tribe to tribe as wolves were from region to region.)

Now it's very unlikely that dog breeding thousands of years ago was anything other than very random. There was no scientific breeding in the modern sense but because dog populations were isolated, the breeding, genetically speaking, was 'clean'. Our intervention in dog breeding, however, from our earliest unintentional intervention, intensified the diversity of both the form and the behaviour of dogs. We produced dwarf and giant dogs, dogs with long noses or squashed faces, erect or lop ears, straight or curly, long or short, black or white hair. About the only physical characteristic retained in all breeds of dogs (other than a leg at each corner) is the classic large chest and small waist.

Our intervention in natural selection allowed this great diversity of physical traits to survive but just as importantly – more so in fact – we also allowed as great a range of behavioural traits to evolve. As I mentioned in the introduction, we accentuated wolf characteristics in our dogs producing terriers that are more aggressive, hounds that scent better. We improved some wolf traits but in doing so we sacrificed others. Domestication of the dog meant that they could count on us for help. Life became 'easier' and dogs could change both physically and mentally but still survive and reproduce. The wolf is, after all, the ideal dog for survival in adverse conditions. How then, did we alter the wolf's behaviour to produce the dog's mind?

There are basic laws of genetics that control the inheritance of many, if not most, canine characteristics and it is these that we employed, although until very recently we didn't actually know exactly what we were doing.

Every living cell contains a data bank, information storage of the most magnificent simplicity but stupendous capacity. The parts of the cell that house the data bank are the chromosomes and these are what are primarily dictating the behaviour of my dogs right now.

Chromosomes are just large enough to be seen under the magnification of a standard microscope, but each chromosome in turn consists of thousands of genes threaded out in a very specific manner, just like beads on a string. When cells divide and reproduce themselves, these genes do too, reproducing in the most amazingly exact way, so exact in fact that geneticists calculate that there might be one mistake in one single gene for every million copies. Compared for example to what I have written in this chapter, that would mean that if the chapter were retyped a million times, only one letter of one word would be spelt wrong once!

Usually, when we think of genetics, we think of its effect on the appearance of the dog, on his coat colour for example. But genetics also controls the dog's mind – how he thinks. Through genetic manipulation, we have changed the dingo to the spaniel. This isn't merely speculation. Selection experiments in rats have shown that natural populations carry numerous genes which affect behaviour QUANTITATIVELY. Rats have been bred to be 'maze-bright' and 'maze-dull' for example. The 'maze-bright' rats didn't have superior minds to other rats though. They were no more intelligent in any other way, only better at finding their way through mazes. Similar experiments were carried out in the 1960s with dogs, and there too it was possible to breed for 'maze-bright' animals.

We, of course, don't need scientific experiments to tell us this. Retrievers are terrific at retrieving and carrying. My older retriever, Liberty, instantly found the apple that had just dropped and brought it into the kitchen for me. But she can't unlock the back door without my help!

In other experiments, mice have been bred with a strong preference for alcohol over tap water, a behaviour that many humans feel might be genetic in them too, but which is a fairly pointless and dead-end behaviour for the mouse. Other mice have been bred with a preference for saccharin over sugar (while taste experiments in dogs show the exact opposite – dogs are exquisitely sensitive at differentiating saccharin from sugar, disliking the former and adoring the latter. I'll explain that further in Chapter Three).

The genetic influence on behaviour is more difficult to understand than the genetic influence on morphology quite simply because there are many genes responsible for behaviour while coat colour, for example, might be under the control of only a few. Genes affect the way the brain works (see the next chapter) which in turn affects the production of hormones (see Chapter Four). It is all these factors taken in composition that affect the dog's mind. On top of this is the influence of the environment, the dog's mother, its littermates and us (see Chapters Six and Seven). There is really no division between nature and nurture. The actual concept of nature and nurture is far too simplistic for reality. Geneticists have created a mathematical formula to determine the heritability of a characteristic, a formula that is great to play with on computers but otherwise of little importance to us.

In the age of computers, however, it might be easiest to visualize the genetics of the dog's mind by thinking of genes as individual data banks. Modern information technology is in fact trying in a primitive way to do what genes already do so perfectly. Genetics is information technology on a vast almost incomprehensible scale.

Rather than simply looking at the effect of a dog's parents on his mind, we could, if we wanted to, go back almost to the beginning of time. Richard Dawkins in his exhilarating look at genetics in *The Blind Watchmaker* describes how cows and peas have an almost identical gene, a gene of 306 characters (I'll explain characters in a minute) where only two characters are different between the cow's gene and the pea's gene. This gene goes back to the common ancestor of the cow and the pea and, of course, of you and me and our dogs. And over perhaps 1.5 billion years, 304 of the 306 characters in this chromosome have remained the same. I've mentioned this, not just because it's an arcane piece of trivia, but to remind us of the bedrock solid base of dog behaviour. The dog's mind is the wolf's mind and the stability of that base is overwhelming. Aubrey Manning, in his pioneering look at animal behaviour published in 1972, said it as simply as possible. 'It is a remarkable testimony to the stability of fixed action patterns that some 8000 years of domestication with, at times, intense selection, has produced so little modification of ancestral patterns.' Qualitatively speaking, genetic manipulation has virtually no effect on behaviour. I cannot produce dogs that can unlock kitchen doors. I can, however, breed selectively to produce dogs that bring food back to the den more frequently – retrievers. Through genetics, we can quantitatively influence behaviour but no more than that. We can make dogs less aggressive, or more maternal, but we can't teach them to understand abstract thought or conditional sentences. 'If you do that again, I'll be very angry!' will only ever be a tone of voice or a body gesture to a dog – no more.

There are sound chemical reasons why fixed action patterns in dogs, or instinct if you prefer, are so stable. Dogs inherit an equal number of chromosomes, 38, from each parent and on each chromosome are genes that are responsible for specific form and function. (It may be that certain genes are ultimately responsible for certain behaviours too. In us, there is a link between a gene and a biochemical imbalance called phenylketonuria that causes mental retardation.)

The gene itself is a superb information storage medium. Modern information technology uses two sites or states, O and A for example, as its units of storage. By creating long strings of various combinations of O and A, information can be stored in a computer's memory.

Genes, on the other hand, use four sites or states, usually called A, T. C and G, for its data bank, and it is obvious that the random combinations of four units arithmetically offers a far greater degree of complexity and sophistication than the two that our computers use.

The common scientific name for a chain is a 'polymer' and many chemicals can be 'polymerized'. Sugars, for example, can be polymerized as starch or cellulose. Polymers can be made in infinite lengths which can, of course, be very useful. Trees, for example, are primarily made of cellulose fibre, infinitely long chains of polymerized sugars.

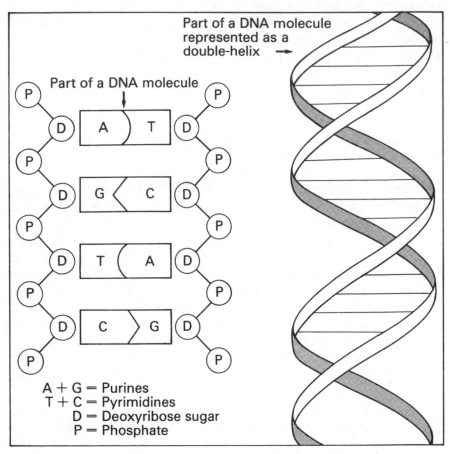

A gene is simply a four-state information technology chain.

The chemicals that makes up genes, A, T, C and G, are called nucleotides and because they come in four-state information technology chains, they are called polynucleotides – a medium of almost infinite capacity. There is enough information capacity in a single cell from either of my dogs to store all the information of all the books that I have read in order to write this one!

The polynucleotide that passes information on from one generation to the next is called deoxyribonucleic acid or more commonly DNA. This is the information that is seared into the memory of germ cells, eggs and sperm, then read millions of times over as cells multiply to first produce the pup, then to keep it going into adulthood. The nucleotides are the characters in the gene's chain, and copying is so amazingly faithful that the cow and the pea, as we have seen, have one gene that goes back almost to the beginning of time.

How then, does each new pup differ from the others? If the replication system is so superb, how can one dog's mind differ from another's? The answer is in the germ cells, the egg and sperm. On each of the 38 chromosomes there are specific 'locations' or 'sites' or 'addresses' that handle certain aspects of life – hair colour or phenylketone production, for example. The 'contents' at each site will vary, but the site itself will be the same in that specific chromosome in every cell of that species of animal. This is the basis for genetic engineering. By knowing what a 'site' is responsible for, it is possible to splice a gene of known function into that site and alter the behaviour of a cell.

All chromosomes come in pairs. In other words, all sites are duplicated in each and every cell, all cells that is, except the egg and the sperm. These cells only have single chromosomes and it is the intermingling and coupling of gene sites from both parents, gene sites with the same addresses but differing contents, that allows for the emergence of different behaviour patterns in pups.

Most genes are amazingly successful at remaining in the archives – the genes that control instinct or fixed behaviours like hunger, thirst, mothering. Even dog coat colours like merle Collies or spotted Dalmatians aren't man made alterations. All are inherited from the wolf's genetic potential. Wolves come in white, blond, cream, red, ochre, grey, slate blue, brown, black and the shades in between. One Eskimo remembers once seeing a spotted wolf, black and white on the tundra of northern Canada.

The natural environment affects genetics through natural selection, in other words, the survival of the fittest. Short ears are less sensitive to the cold and are sensible on dogs in the north. Long ears dissipate heat and are more practical in southern climates. Every now and then a mutant gene, one change in one letter in a billion letter text, will occur and will perpetuate a certain characteristic such as ear size.

Mutations don't just happen. They're caused by x-rays, cosmic rays, radioactive substances, chemicals or even other genes called mutator genes. Tay-Sachs disease, a disease of the nervous system that only affects Jews of European origin, probably has its origin in a genetic change caused by a cosmic ray hitting someone (who happened to be Jewish) in Eastern Europe in the 1400s. Huntington's Chorea has a similar one in a billion origin.

These relatively minor mutations may continue to exist and 'flow' from generation to generation and through our intervention in dog breeding, certain ones will 'cluster' in some breeds. A mutation long ago, that caused an animal not to chew and swallow its food but to carry it back to the den for its young, ultimately helped make she wolves good mothers and made it more likely that the species would grow strong and survive. This characteristic is dispersed among all dogs today but is 'clustered' in the retrievers which is why Liberty brought the fallen apple to me this morning.

There is a good and a bad side to this clustering of mutations from natural behaviour. Natural selection does NOT create superdogs. On the contrary, it creates median dogs, middling dogs. It is human intervention in canine genetics that enhances and creates 'supertraits'. In that sense, there is no logic to the suggestion that dogs have become degenerate simply because they are no longer adapted to live in a wild environment. Just the opposite can be argued, that dingos and similar wild dogs are degenerate because they haven't evolved to live in a modern environment.

The drawback to 'inbreeding', the breeding of related animals to accentuate certain traits, is that while improving the dog in one way, by uncovering latent qualities such as retrieving or pointing, there can also be an accompanying accumulation of genetically linked unwanted and deleterious effects. The Boxer dog might be a good guard and fighter but as a breed, it suffers from skin cancer more than any other. Cancer has been accidently but genetically selected. On more of a behavioural line, pointers have been selectively bred for gentleness, docility and freezing on sighting game, but sometimes this 'hothouse' breeding goes too far and produces neurotic dogs, Allegheny Sues. And if Allegheny Sue had been a champion show dog, her influence would have been much worse. A champion that carries a deleterious trait can easily spread it through the breed. This is most obvious with medical problems such as progressive retinal atrophy, an inherited blindness in some breeds of dogs, but behavioural problems can be spread through a breed in the same way. Just selecting for coat colour might accidently do it! Cocker spaniels can suffer from a behavioural condition called 'avalanche of rage' syndrome or 'Jekyll & Hyde' syndrome, a behaviour where the dog, for no apparent reason, quite suddenly becomes fiercely aggressive and then, just as suddenly, reverts to a happy-go-lucky demeanour. Curiously, this problem occurs most frequently in blond or golden

Cockers, less frequently in solid black Cockers, and in those of mixed colours hardly ever at all.

Dr. I. Stur of the University of Vienna Animal Genetics Institute has gone on record as saying this about dogs. 'The more beautiful a dog is, the worse is its constitution.'

'I'm quite certain that Dr Stur was trying to be provocative but he bases his statement on research carried out in Austria. There he investigated performance traits of 62 German Longhaired Pointers. The 21 different conformation and temperament characteristics were classified by experienced dog judges.

Dogs were judged for:

pointing manner	form value	hair quality
constitution	eye characteristics	high nose
deep nose	hardness	speed
temperament	warm tolerance	hunting passion
barking	scent work	shot proofness
length	height	teeth
bone thickness		

As you can see, this list includes both objective and subjective qualities. What Dr Stur observed was that, according to the 'unbiased' judging of his judges, there was a negative correlation between 'form value' and 'constitution'. 'Form value' in fact means breed standards. Stur argues that by selectively breeding only to morphological breed standards, we unknowingly breed for similar genes at gene sites or 'homozygosity'. And unwittingly, as we breed for homozygosity to conform to breed standards, we also breed for homozygosity in the genes that affect behaviour.

Stur's research showed that pointers, which were more likely to win at

Some dogs excel in the show ring.

dog shows, were less likely to function in the field as pointers. That was 's negative correlation between beauty and constitution. Specifically, he says there was a negative correlation between shot proofness (the opposite of gun shy) and hunting passion. He went further and suggested that shot proofness could be used as an indicator of the calmer and more obedient dog and could be a worthwhile factor to include in judging whether a pointer would make a good pet.

Stur's point has already been acknowledged in many other working breeds of dogs. There is a distinct difference in the temperament of the show golden retriever and the working lines in that breed. Show lines can be more flighty, even aggressive, while working lines retain the calmness of nature that originally made them good field dogs.

Others fail in the show ring but excel in field work.

Selective breeding is a precariously balanced act that brings together a highly selective gathering of genes from the general pool of dog genes. Selective breeding appears to influence the threshold of certain behaviour patterns. A lower threshold for a certain behaviour allows us to bring out and shape that innate behaviour through training. That's how we have created dogs that are willing to live together in large groups but not fight as much as would normally be expected (beagles), or the opposite, dogs that if kept in groups of four or more will invariably fight (wire haired fox terriers). Selective breeding, and that in most cases usually means breeding relatives, simply makes certain innate behaviour patterns more accessible – guarding, herding, leading, guiding, hearing, retrieving. In breeding for these traits we alter structures too.

The most obvious problems of breeding close relatives for a few generations are the deleterious physical traits that ensue. Scott and Fuller in the most comprehensive experiments ever carried out on the influence of genetics on dog behaviour created Basenjis with undershot jaws and Shelties with a tendency to fat in only a few generations.

Professional breeders often understand this, which is why they can appear ruthless in culling new litters. Our intervention in canine genetics has dramatically increased the death rate in newborn pups compared to wolf pups, although the dog's increased fertility compared to the wolf compensates for these 'new' losses. The geneticist M. B. Willis followed 52 litters of Briard pups and recorded that 6.4 per cent died naturally within 48 hours of birth and a further 12 per cent were culled deliberately. In figures, this means that one in five pups is lost and the effect on the genetics of the breed is obviously considerable.

There is no question that through genetic manipulation we can modify the dog's mind. We can increase the dog's friendliness to us, its ability to freeze, its physical stamina, its courage in the presence of other animals, other dogs or us. Service organizations use this knowledge in producing 'new' dogs. In Britain, the Guide Dogs for the Blind organization now has the largest breeding programme in the world, producing golden retriever – Labrador retriever crosses, genetically selected for training as seeing-eye dogs. In South Africa, the paramilitary police have crossbred Bloodhounds with Rottweilers as well as Dobermanns to produce dogs with the scenting and tracking ability of the former and the constitution of the latter.

Theoretically, undesirable behaviours in dogs can be altered or avoided by careful attention to genetic influence but, in fact, we still breed primarily for morphology and often simply for curiosity. By doing so, we retain within the genetic repertoire of the dog, a genetic pool of what can only be described as harmful recessive genes, genes that might on one hand create the morphology that we desire, but that on the other hand might be linked with a deterioration in maternal care or a carelessness in hygiene or an increased tendency to snap at children. Theoretically, we should be able to breed for the future, for example to breed for dogs with exemplary cleanliness, dogs that are 'house-proud', dogs that are affable with children but protect their homes against unwanted strangers. A study of Swedish Army dogs, however, suggested that mental states like affability or disposition to defend have low heritability values. In other words, it might be easier to use other influences – early learning for example – to alter the dog's mind in these ways. The primary wiring, however, rests in the inherited data bank – in the genes. The most important fact that genetics tells us is that rather than look at the potential of the dog's mind we should be aware of the perfection of the system and the immutability of change. This means that in examining the dog's mind, we should always remember that no matter how we manipulate the dog, it will be limited to its inherited repertoire from the wolf. If we understand the limitations of change, we can be more successful in influencing the dog's mind.

Chapter Two

The Brain

The brain is the body's great synthesizer. It needs vast amounts of fuel in order to function properly. Although the brain of the average dog accounts for less than half of one per cent of his body weight, it receives over 20 per cent of the blood that is pumped out by the heart. It's the blood-greediest organ in the dog's body – a furnace of activity.

The dog's brain is, of course, responsible for interpreting and acting upon all the information or signals it is sent by the senses and by the body's hormones. The dog's response to these signals has been, to a considerable extent, predetermined by the 'fixed wiring' of its genetic make-up, but that does not mean that he can only respond in a consistent or mechanical way. The dog's brain does not have a reflex-like in-flexibility as some behaviourists infer. When an apple dropped from the tree onto my dogs, one ran in panic while the other picked it up and brought it to me. They both received the same signal but acted differently. Their brains synthesized the information that was received, and based on their genetic prewiring but also on past experience, the quantity and the quality of their responses varied.

There are two ways in which the brain stores information. It stores information concerning the relationship of one event to another. When Pavlov's dogs heard a bell each time they were fed, they soon learned to salivate simply at the sound of the bell. They were 'conditioned' to respond to the bell's ring by salivating, and this is called the brain's 'conditioned response'. It is also sometimes called 'stimulus expectancy'.

The other way that the brain stores information is called 'instrumental conditioning' or sometimes 'response outcome expectancy'. My dog bringing the fallen apple to me is a result of instrumental conditioning. She has been trained to retrieve and, just as Skinner's rats learned that if they pressed a lever food would appear, so too my dog has learned that if she retrieves for me she gets a pat on the head. I'll discuss these different learning methods in Chapter Seven.

Both of these responses depend upon the genetic potential of the dog as a species and the individual dog's information storage system, the brain. In other words, his response depends upon the actual circuitry, the wiring of his brain. This is an area that is so densely complicated that we really have very little understanding of the actual anatomy.

The dog's behaviour is plastic; it's malleable and is ultimately stored at the cellular level of the brain. A behaviour, running away from a fallen apple, or bringing it to the back door, involves networks of brain cells intercommunicating with each other and acting upon information that has been received by the senses. Brain cells themselves are called neurons and each one consists of a cell body, together with a thread-like axon at the end of which there are receptor sites that make contact with other neurons. The receptor site is called a synaptic bulb and a typical neuron has several thousand synaptic bulbs.

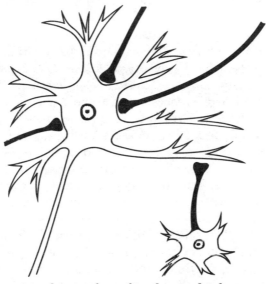

A neuron showing four of its thousands of synapses.

Some neurons have over 10,000 synaptic bulbs. That means that one nerve cell can send different messages to thousands or tens of thousands of different cells simultaneously. The magnitude of the potential of the brain becomes even more mind boggling when you consider that there are literally billions of cells in the dog's brain. (We humans have about 15 billion cells.) Consider what a simple bee can do – carry out selective hive duties, estimate the sun's angle to within three degrees, know the time of day to within half an hour, estimate the weight of pollen it's carrying, steer courses according to landmarks, measure dance movements of other bees and determine how far away food is, recognize and attack enemies, and all with a brain less than three quarters of a cubic centimetre. If a bee can do that, then imagine the complexity of the neuronal connections of the dog's brain.

When I was at university, it was thought that there were only two chemicals that acted as transmitter substances at nerve synapses, acetylcholine and noradrenaline. It has now been realized that there are at least thirty, and that one nerve cell can actually produce many different neurotransmitter substances. Chemicals called amino acids are probably the major neurotransmitters in the dog's brain and all the chemicals are classified as either excitatory (E) or inhibitory (I) substances. These excite or inhibit other nerve cells. Perhaps the most fascinating of these recently discovered substances are the encephalins or endorphins, the body's own opiates, that are manufactured in the brain and regulate the dog's conscious perception of pain. (Cats have a more powerful endorphin system which is why they are better at coping with physical pain.)

The speed of transmission of information from brain cell to brain cell and throughout the dog's nervous system depends upon whether nerves have developed a fatty substance called myelin to protect them, and here is one of the dramatic differences between the dog's brain and ours. Our brains are immature at birth and continue growing for years after. There is, for example, a sixfold increase in the surface area of the human brain by six years of age and the frontal lobes (I'll explain the parts of the brain shortly) continue growing until ten years. Myelination of the human brain continues until the end of the third year and in fact there are important areas of the cortex, the association areas, that are not completely myelinated until eighteen years of age.

Myelination and maturation of the dog's brain is infinitely faster. Like us, there is virtually no myelination at birth. An unmyelinated nerve transmits impulses at around 2 metres per second (4.5 mph) which is why a pup's response to almost any type of stimulus is so slow. Pups certainly do have less sensation and this is the reason that tails are normally amputated in certain breeds when they are less than three days old. It is wrong, however, to say that they don't feel pain when this cosmetic procedure is carried out. It might appear that they don't feel pain, but this is due to the delay in their response because their nerves aren't myelinated yet. A large myelinated nerve fibre transmits impulses at 120 metres per second (270 mph), sixty times as fast, which is the reason why tails can't easily be docked later.

At birth, the dog's brain is only sufficiently developed to control the heartbeat, breathing, balance and equilibrium, and probably has a dramatic ability to repair itself. As an analogy, if the part of the brain that is responsible for registering sight, the visual cortex, is removed from an adult cat, as has been done experimentally, that cat is blind. But if the same area is removed from a newborn kitten, it will be able to see as an adult. The same is probably true with the newborn pup's brain, which has an ability to reorganize itself amazingly even after early severe damage. The reason for this is that just before birth there are far more cells in the brain than at birth, and these cells have more widespread connections. The genetic programme that is responsible for the development of the brain plays safe by building in extra cells and connections. This genetic programme encourages the brain to form exploratory networks that are ultimately sculpted into the brilliantly functioning synthesizer that the brain becomes. Death, it seems, plays a great part in the development of the brain but the results of this unpleasant cat experiment also show that the development of the brain is influenced by experience, something I will come back to.

The newborn pup has no temperature control but his facial nerve is mature and he uses his head as a probe, avoiding cold hard surfaces and liking soft warm ones. He already has a sense of smell however. If aniseed oil has been placed on one of his mother's nipples and he sucks from it, he will later crawl to a cotton bud soaked in aniseed oil when he is cold or hungry.

By two to three weeks of age, his brain is sufficiently developed to control his body temperatures and his metabolism. He can, for example, urinate and defecate without requiring his mother to stimulate these activities by licking the anogenital region. By four weeks of age, he has conscious perception of space, of what part of his body is being touched. By five weeks of age, he has the ability to stay awake if he wishes to do so. This means that the reticular formation, a system of nerve cells in the hind region of his brain that among its functions is responsible for sleep, has developed.

Sleep is a good example of how a behaviour can be affected by neurotransmitter substances. In the brain there is a sleep centre (the nerve cells of which produce an inhibitory (I) substance) and a waking centre. In 1957, W. Dement discovered that there are really two sleep centres in the brain, one responsible for deep sleep and the other for light sleep. The deep sleep centre in the hind brain is there from birth, but the light sleep centre in the forebrain doesn't develop in the dog until he is four to five weeks old. Dement called deep sleep 'activated sleep' and correctly associated it with REMs, rapid eye movements, the visible sign of dreams. It's assumed that when dogs have REMs, when their eyes move beneath their lids, when they paddle with their feet or twitch their lips and noses or sometimes even bark, that they are dreaming, are in activated sleep.

In addition to the two sleep centres, there is the waking centre, also under the control of chemical substances. If this centre temporarily malfunctions, the dog falls into a deep sleep, REM sleep. It might only last for seconds but it is true sleep. The condition is called narcolepsy and is known to occur only in four breeds, the Dobermann, the Poodle, the Dachshund and the Labrador Retriever. Narcolepsy is a classic example of how genes control the chemical functioning of the brain and how chemical functions are ultimately responsible for the dog's behaviour.

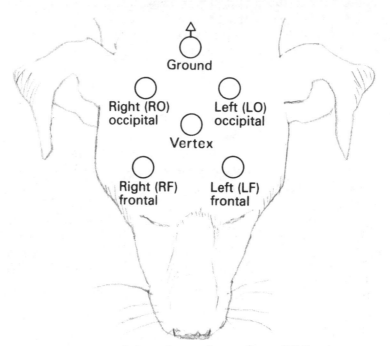

Points of electrode attachment for an EEG.

Many of these early brain changes can be monitored by using an electroencephalograph (EEG), a machine that measures changes, oscillations, in the electrical potential of the brain. We know, for example, that there is little activity in the pup's cortex during its first weeks of life because there is virtually no electrical oscillation. By three weeks of age, however, medium voltage potentials of a low frequency appear on the EEG tracing and by five weeks of age, the EEG pattern is dramatically similar to that of the adult dog. Anatomically as well, the pup's brain at five weeks of age has most of the appearances of the adult brain. By 22–30 weeks of age, the EEG pattern is identical that of the adult.

Growth of the dog's brain, arborization of the neurons to make their

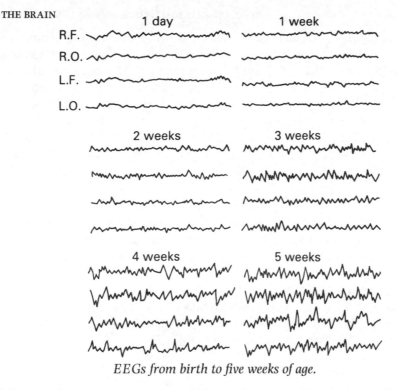

R.F.

R.O.

L.F.

L.O.

1 day   1 week

2 weeks   3 weeks

4 weeks   5 weeks

EEGs from birth to five weeks of age.

billions of synaptic contacts and myelination of the nerve fibres is complete in half a year compared to eighteen years in humans.

Scientists have for years tried to develop a method of comparing intelligence between one species and another. Harry Jerison, an American authority on the evolution of the brain, used the speed of maturation of the brain in his mathematical model to compare intelligence, and developed the encephalization quotient, the EQ, to compare brains from different species. Using Jerison's logarithmic calculations, carnivores have higher EQs than do herbivores. His is a neat mathematical formula and it tries to take into consideration the fact that the larger an animal's body, the larger the brain must be to control it, but the EQ still misses the point that different species have different lifespans and consequently will have different rates of maturation of their brains.

Others have tried to compare brain function in various species by looking at comparative maze learning. Ron Kilgour at the Ruakura Animal Research Station in New Zealand did this and observed that only children were better than dogs at this activity.

Children
Dogs
Cows

Goats
Ewes
Cats
Rats – Wistar strain
Ferrets
Hens
Pigeons
Guinea pigs
Mice
Brush tailed opossum

But we have already seen that rats can be selectively bred to be 'maze bright'. If Kilgour had used a strain of maze bright rats, they would probably have scored fewer errors than dogs. Comparative maze learning is really a pointless test for comparing brain function between species.

The anatomy of the dog's brain is similar to that of most other mammals. Classically, the cerebrum controls learning, emotions and behaviour, the cerebellum controls the muscles, and the brain stems connects to the peripheral nervous system.

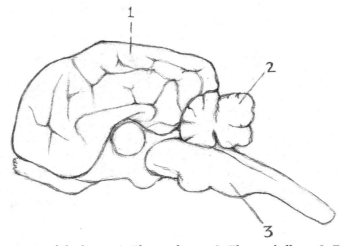

The three parts of the brain: 1. The cerebrum, 2. The cerebellum, 3. The brain stem.

It's not quite that simple however. The cerebral cortex is indeed important for sensory discrimination and sensory discrimination, as we shall see in the next chapter, is of paramount importance to the functioning of the dog's mind. The cortex is not as important as it was once thought to be for learning and problem solving. Another network called the

limbic system is probably the area that controls general memory functions with the hippocampus as perhaps its most important part. A dog
understands his own relationship to the world around him through the
proper functioning of his limbic system. This is important when it comes
to our influence over the dog's behaviour. The dog is motivated by
different reasons to those we experience, and the conflict between what
he 'instinctively' wants to do and what we want him to do is played out
in the limbic system of his brain. If we can override this system, either
by increasing the reward he gets for obeying us rather than himself, or
by punishing him for his 'normal' behaviour, we can control his
activity.

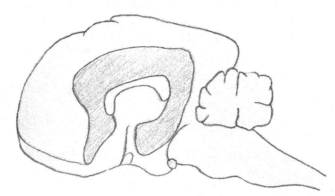

The conflict between what a dog instinctively wants to do and what you want
him to do is played out in the part of the brain called the hippocampus.

What we need to remember about the various parts of the brain is that
virtually none functions on its own. The dog's brain is a complex
network and follows the general rule that the more complex a brain is,
the more variable are the mind's behavioural responses. When the
external event is recorded by one of the dog's senses – when my dogs
saw or heard the apple fall to the ground – that information was passed
from the eyes and ears simultaneously to the cerebral cortex, to the
limbic system and to the reticular formation.

Because the apple dropping event was significant to both dogs (another
dog might not have noticed or cared), their reticular formations maintained brain involvement while the limbic system decided how the
event was going to affect the dog. My younger dog's limbic system
registered surprise, fear, perhaps even pain and was ultimately responsible for her running fearfully to the kitchen door. My older dog's limbic
system recorded curiosity and led to her picking up the apple and
bringing it to me. It did so this way.

The dog's limbic system, and specifically the hippocampus (so called because in our brains this coil shaped structure is reminiscent of a seahorse), is responsible for maintaining a value system in the dog's mind. When the dog is confronted with a specific situation, any discrepancies between the value system's expected outcome and what really happens are noted in the limbic system. It does this through its connections with learning centres in the cerebral cortex and through other connections to the hypothalamus, the part of the brain which in turn feeds into the body's hormone system. If the reward we are offering is less rewarding than what the dog is engaged in, it is in the limbic system that he 'decides' not to obey a command. It's interesting that the rabies virus in dogs can cause two dramatically different behavioural changes. The dog can become vicious or, just as frequently, suffer from 'dumb' rabies. On post mortem examination, the evidence of rabies in the dog is the presence of substances called Negri bodies in the brain. Negri bodies have a natural predilection for the limbic system and are always first looked for in the dog's hippocampus.

One of the fascinating aspects of the dog's mind is that learning improves when the emotion of INTEREST is aroused, an emotion that the hippocampus is involved in. The hippocampus is probably the site of instinctive behaviour but it is also responsible for new memory, though not for old. Old memory is stored in the cerebral cortex.

From the viewpoint of behaviour, the part of the cerebral cortex that is most interesting is the neocortex and specifically that part of the neocortex called the association cortex. This is the region of the brain that is responsible for receiving items of information passed on from the limbic system, determining the relative importance of the information, comparing the information with previous experience, selecting a suitable response to this information and predicting the consequences. It does so by sending chemical messages across synapses at 120 metres per second through that vast meshwork of billions of cells with no beginning and no end.

The dog's association cortex works in exactly the same way that ours does. The only difference is in the quantity and advanced degree of sophistication of our cortex. Learning through conditioned reflex occurs in the association cortex. Although Pavlov didn't know it, that is the part of the dog's brain that made them salivate when they heard the bell ring.

The parietal and occipital lobes of the cerebral cortex are the regions that are involved in a conscious recognition of what is happening, but the temporal lobes are concerned with learning and with memory. This

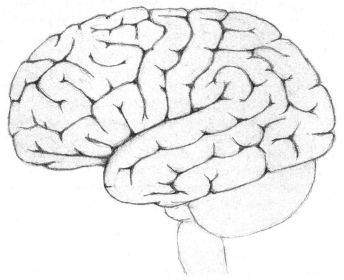

Human cerebral cortex.

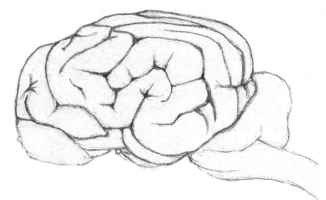

Dog cerebral cortex.
The only difference between the cerebral cortex of the human and the dog is
the quantity and advanced degree of sophistication of the human cortex.

is the storehouse of past memory. When a dog shivers in trepidation as it enters my veterinary clinic, it does so because the temporal lobe of its cerebral cortex has released some previous information about me or my clinic. When an Inuit's Husky dog shivers and digs into the snow to build itself an ice house however, it is the dog's limbic system and connections to the hypothalamus that are controlling his behaviour.

The temporal lobe is also connected to the hippocampus and through it to the hypothalamus and on to the body's hormone system. This is why, when the dog finally sees me, it might try to either fight or flee. It might even urinate in panic, a sure sign of an activated hormone system. I will discuss the hormonal feedback on my frightened patient and the cold Husky in Chapter Four.

We know that the temporal lobes of the cerebral cortex are where memory is stored as a result of work done early this century by the Canadian surgeon Wilder Penfield. Penfield stimulated the temporal lobes of conscious human patients and they recalled hearing a tune or seeing some past event. More recently the neurophysician Oliver Sacks in his book, *The Man Who Mistook His Wife For A Hat*, recounted a case in which one of his patients, an elderly woman in an old folks' home in New York, suddenly started hearing Irish folk tunes. It sounded to her as if someone had left on a radio that played the same four tunes over and over again. Sacks explained that in fact this patient, a woman of Irish descent who had emigrated to America when she was less than five years old, had had a stroke and that the stroke had 'released' these old songs from the memory storage data bank in her cortex. What this demonstrates is that ALL past significant experience is probably stored in the memory centres of the brain and this is why dogs are so good at recognizing people or other dogs after years of separation.

The temporal lobe's connection to the hippocampus and hypothalamus might also explain the relationship between memory and the emotion INTEREST. All good dog trainers know that a dog must be interested in training otherwise the exercise is hopeless.

The greatest difference between the human brain and the dog's brain is in the size of the frontal lobes of the cerebral cortex. These are the areas involved in human intellectual function. I'm using my frontal lobes to try to explain brain function in a simple way. This is something that neither of my dogs will ever be able to do, no matter how well they are trained, simply because they haven't got big enough frontal lobes. Dogs DO have frontal lobes and they are responsible for alertness, intelligence and the temperament of the individual dog. Damage to this part of a dog's brain results in stupidity, inactivity or sometimes viciousness and hyperactivity. Curiously, removing the frontal lobes of the brain results in the dog tolerating problems and conflicts better but he also has a reduced ability to cope with his environment. At one time frontal lobotomies were actually considered as a treatment to overcome canine behavioural problems. The procedure has no ethical, moral or medical justification. Nor for that matter do the invasive procedures that are still used by some research workers in their search for the brain's learning centres. Cutting or burning bits out of animals' brains is primitive, ethically wrong and scientifically unjustified. The brain is simply too complex to study this way. There are too many interconnected loops and tracks. Remember, almost every part of the cell network of the brain interrelates to other parts.

In simple terms, the brain is an almost infinitely complex data handler and there are no simple scientific or philosophical answers to its functions. If we are to learn anything about the dog's mind by studying the dog's brain, it is this. If a young dog is 'superstimulated', if it hears loud noises, sees flashing lights, has to balance itself on an inclined plane for a few minutes each day when he is very young, he will grow up to have a larger brain with more cells, bigger cells and more interconnections between them. It isn't just a matter of genetics, of instinct, of set pattern functioning. If a dog is deprived of stimulation, he will have a smaller brain. We can influence the development of the dog's brain, we can influence his mind by providing him with the best environment possible when he is a new born pup. Finally, mongrel dogs have larger brains relative to their body size than do purebred greyhounds. Is this because it's more likely that mongrel dogs will experience a broader range of sensory stimulation when they are very young or is it a manifestation of genetics – of hybrid vigour? As yet there is no answer.

Chapter Three

The Senses

THE INFORMATION THAT FEEDS THE MIND

To understand the dog's mind, we have to enter a different sensory world -- the world of canine sensation -- and imagine how the dog perceives and responds to life around him. This is not difficult to do with respect to some senses. It's not hard to imagine seeing in shades of grey as a dog does because we have all had the experience of watching black and white television or films. For other senses though, it's almost impossible. How can we ever appreciate the awesome capacity of the dog to differentiate thousands of different odours, some as dilute as one part per million? We humans can only really use our own capacities for terms of reference and it is this massive restriction that inhibits us from really understanding exactly how sensory input, touching, tasting, hearing, seeing and smelling, affects the dog's behaviour. There are, however, some good anatomical clues and experimental evidence that can help us more fully understand how the senses affect the dog's mind.

We already know that, quantitatively, all dogs share the same genetic potential. They all have the same information technology and are wired up to the same hardware. The Saluki and the Beagle both have the same quantitative capacities -- they hear, see, touch, taste and smell in the same ways. Qualitatively, however, there are considerable differences. The Saluki's vision is superior to the Beagle's. The Beagle on the other hand has a massively superior capacity to follow scent. Qualitative sensory capacity is one of the dynamic differences between different dog's minds.

TOUCH

A two-day-old pup, when he is separated from his mother, will cry and swing his head back and forth like a pendulum until he touches his mother's body. Then he stops crying and crawls to her. Touch is the earliest and possibly the most important of all the canine senses.

Michael Fox has reported that Dr Y. Zotterman of the Swedish Research Council discovered infra-red receptors in the dog's nose and it is these that lead the still blind and deaf newborn pup back to its mother, but it is touch – contact comfort – that soothes the pup and is overwhelmingly important for the development of a mature and sensible mind.

Harry Harlow's experiments at the University of Wisconsin in the late 1950s with newborn macaque monkeys revealed just how distressing it can be to be deprived of touch. Baby macaques, that were isolated from birth from their mothers and from other young monkeys, grew up to suffer from overwhelming and serious behaviour disturbances. They crouched motionless all day or simply clasped themselves. They rocked their bodies in a monotonous and stereotyped way, banged their heads against the walls of the room and mutilated themselves. Touch is the primal sense in dogs as it is in us. It is more important than any other sense for the development of the normal adult, and dogs that are deprived of touch will grow to become subordinate, fearful and withdrawn.

The importance of touch continues throughout the dog's life. Touch remains forever the most potent reward that a dog can receive, more important than even food, unless the dog is trained with food rewards. Stroking a mature dog that knows you can reduce his heart rate, lower his blood pressure and drop his skin temperature. In other words, stroking reduces the dog's state of arousal. (The same thing happens to us too. Stroke a dog to which you have formed an attachment and your state of arousal also diminishes.)

As well as its importance for emotional well-being, touch is also used by dogs to investigate their environment. In common with many other mammals but not us, dogs have special sensory hairs, vibrissae, above their eyes, below their jaws and, most importantly, on their muzzles. These sensory hairs are imbedded in areas of skin that have intense blood supplies and numerous nerve endings. Dogs can sense air flow and current with their vibrissae as well as determine the shape and texture of objects. Of course, the dog's entire body has sensory nerve endings that are stimulated by touch. These can sometimes respond in an abnormal way and cause a common behavioural disorder.

There is a skin condition in dogs called acral lick dermatitis. Dogs that suffer from this complaint can obsessively lick their legs, usually their forelegs, until they have licked themselves raw. It's one of the more frustrating problems in veterinary dermatology and its cause had been assumed to be boredom. However, an article in the Journal of the American Animal Hospital Association suggested that lick dermatitis could in fact be a peripheral sensory nerve disorder. The researchers observed that dogs that suffered from lick dermatitis had lower amplitude evoked nerve action potentials in the touch sensitive nerve endings of their forelegs. They concluded that the behaviour of these obsessive lickers was related to impaired touch sensitivity to their legs.

TASTE

The only other sense that is functioning at birth is taste (simply because the only parts of the brain involved in sensory function that are myelinated at birth are those associated with touch and taste.)

The sensation of taste is closely associated with smell and it can be difficult to differentiate between these two 'chemical' senses. Dogs don't have as refined a taste sense as we do and this is in part an anatomical difference. We humans have around 9000 taste buds on our tongues whereas a dog has, according to one researcher, 1706. Most of these are on the anterior portion of the tongue. It has always been assumed that dogs share the human taste world and, although they have taste buds like us that register sweet, sour, bitter and salty, they do so in such a unique way that it is probably better to discuss a dog's taste sensation as 'pleasant – indifferent – unpleasant'. For example, unless dogs are trained to eat sweets, they are rather indifferent to them, unlike rabbits and horses, both of which have a genuine 'sweet tooth'. Dogs do, however, prefer sugar to saccharin in low concentrations in their food and actively reject biscuits with saccharin in them.

In addition to the receptors for sweet, sour, bitter and salty, dogs are also thought to have additional primary taste receptors, like cats and pigs have, that respond to water. This means that dogs might be able to taste different types of water.

Katherine Houpt at Cornell University has researched taste in dogs extensively, doing some of her work for the Committee on Nutrition of the Canadian Veterinary Medical Association. Houpt observed a number of factors concerning taste in dogs.

In the laboratory, dogs preferred canned meat to fresh meat, cooked meat to raw, and meat to cereal. She noted that pet dogs behaved differently. The pet dog's appetite was affected by the taste, texture and smell of the food but also by the pet owner's perception of the pet and the owner's own preferences for food. Finally, the dog's social and physical environment affected its eating habits.

In Britain, it's estimated that at any given time approximately 30 per cent of the dog population, two million dogs, are overweight. There is certainly a breed disposition to obesity and different breeds will naturally

defend different body weights and degrees of fatness. Just compare the
Saluki and the Pug to see how different the normals can be. It is aside
from this normal range that 30 per cent of the canine population allow
themselves, or are allowed, to become fat. How does this happen?

In a kennel environment, dogs that are given 24 hours a day free
access to food eat many small meals mainly during daylight hours and
do not become fat.

One of the reasons for obesity in pet dogs is undoubtably the increased
tastiness of commercial dog food, but there could be more reasons and
these are what Dr Houpt looked for. Many facts are already known.
Huskies in Alaska double their food intake in the winter. Increased
palatability increases eating. So does social competition. Female dogs eat
less when their estrogen hormone level is high. In order to create a
feeling of satisfaction food has to stimulate both the mouth and the
stomach.

Houpt's own research into taste was revealing. Female dogs have a
slightly greater preference for sugared diets than did male dogs and all
dogs preferred their own food warm to cold. And although it is the
odour of food that initially attracts a dog, odour does not play a part
once the dog starts eating. At that point it is up to the taste buds and
texture receptors in the mouth.

Houpt observed that as a general rule dogs preferred canned or semi-
moist food to dry foods but had no preference for canned food over semi-
moist. Smell, not taste, was the important factor in establishing a
preference for one meat over another. In dogs without a sense of smell,
there was no preference for one meat over another but amongst dogs
that could smell, 85 per cent preferred beef to other meats. Dogs that
could not smell, anosmic dogs, still preferred meat to cereal, a good
indication of the importance of taste and texture.

Now Houpt's research gets interesting. Having carried out taste tests
on laboratory dogs in kennels, she repeated these tests on pet dogs in
their own homes and discovered that dogs that slept in bedrooms liked
one food best and dogs that received table scraps liked another. Dogs
attached to the man of the household liked a different food to dogs
attached to the woman of the household. Her conclusion was that taste
alone does not determine a dog's preference for food. There was more
than just sensory input affecting a dog's eating behaviour, more than
just conditioned reflex or instinct. Our interaction with our pets somehow
affects their taste buds.

Pet owners have reported finicky or selective eating habits in female
dogs about twice as often as have owners of male dogs and the answer
would seem to be a hormonal difference. I'll discuss the hormonal

influence on behaviour in the next chapter but the fact that in these eating reports most males are intact and most females are spayed means that the cause of the underlying difference remains unclear.

Of course a great owner variable on the dog's taste buds is human obesity. Ron Anderson at the University of Liverpool has reported that more obese people than lean people have fat dogs. He also reported that older people had fatter dogs than younger people and that dogs fed canned food were less likely to be obese than dogs fed table scraps.

Taste is an important sense in omnivores such as ourselves. In scavenging carnivores like dogs, what is more important is how much they eat rather that what it tastes like. A depletion of calcium and other minerals in the diet can have a significant influence on the dog's mind, producing increased aggressive or exploratory behaviour but as a general rule we should remember that in dogs, palatability is based on the odour of the food first, then its texture and finally its taste.

HEARING

Although the dog's taste buds are not as refined as ours, simply because they don't have as many of them, his hearing is in many ways superior. It has been written that a dog's hearing is four times more acute than ours, a misleading statement for in fact dogs can actually hear sounds from four times the distance we can but still hear the sounds in a broadly similar fashion. Dogs are, however, better at detecting higher notes than we are. The range of hearing for a few species is like this:

Man	up to 20,000 cps
Dog	up to 40,000 cps
Cat	up to 45,000 cps
Bat	30–98,000 cps
Dolphin	100–130,000 cps

In humans the lower end of the sense of hearing merges with the sense of vibration and both dogs and people have similar hearing abilities at this end of the scale. In fact, the upper limit of the dog's hearing is lower than for most other carnivores but, then again, there are few sounds in nature that are so high pitched. One of them is the cry of the bat with a pitch up to around 100,000 cps. We can't hear anywhere near this pitch and apparently Central and South American cattle can't either, which is why vampire bats feed on their blood. Dogs in those regions rarely get attacked by vampire bats, a fact that suggests that dogs might be able to hear some aspect of the bat's cry, and could lead to a wholly unexpected new guard service for dogs, as ears for grazing cattle.

The dog's ability to hear high-pitched sounds is an aspect of the wolf in him. Small prey – mice and voles – make high-pitched squeaks and certainly amongst Canadian timber wolves these small mammals are a main part of the summer diet.

The mobility of their ears allows the dog to scan the environment for sound.

The dogs' hearing is more acute than ours in the 1000–8000 cps range but they also have a distinct advantage over us, mobile ears. The mobility of their ears allows the dog to scan the environment for sound and then to collect the sound waves. This is something that has to be learned. J. A. Altman at the Pavlov Institute of Physiology in Leningrad conducted experiments in dogs that showed they could locate the source of a sound in six hundredths of a second. A dog might use one ear to locate the sound, then both ears to catch the maximum number of sound waves. This is why they can hear over a greater distance than we can but to do so, to cope with the asymmetry of the ear positions when they are localizing a sound, the dog needs a more sophisticated central nervous system computation centre and this is one reason why more of a dog's brain is devoted to sound than is ours. Interestingly the ear volume of the dog increases linearly with weight up to 10Kg but over 11Kg the ear volume doesn't increase with size.

Determining how well a dog hears has been a continuing problem in veterinary medicine. Until recently, all that I and my colleagues have been able to use is the startle reflex – make a loud noise and observe the dog's response. In 1984 at Cambridge University, veterinarians adapted a test called brainstem auditory evoked response (BAER) to test hearing in dogs. Using this test, BAER potentials were not detectable in pups at two weeks of age but were there by three weeks and resembled the adult wave form by seven weeks. No BAER was evoked in dogs that were known to be deaf. This is an excellent method of determining hearing in each ear, the only drawback being the high cost of the apparatus.

Hearing in dogs is in many ways similar to our own hearing so it is relatively easy to understand. The ears catch sound and channel it through to the brain where the significance of the sound is determined. The context of the sound is very important for it is the context that determines the quantity and the quality of the dog's response. There is not a reflex-like inflexibility on the part of the animal's response to sound. Although part of the dog's response to a sound is mechanical, it is not a simple 'lock and key' model of stimulus-response. Remember, when Lexington heard the apple fall, she mechanically ran to me for protection. When Liberty heard the same sound, she retrieved the object.

The acuteness of a dog's hearing ranges over about eight and a half octaves, the same as us (compared to ten octaves over which a cat can hear). Pavlov was the first scientist to investigate pitch discrimination in dogs and showed that a dog can distinguish two notes differing by only one eighth of a tone. That is why some dogs are so good at recognizing the sound of their master's car engine as opposed to anyone else's. It's why dogs can differentiate between a shepherd's whistled commands. Pavlov's dogs salivated at chord differences but not in between. Do dogs then have absolute pitch? Are dogs musical? The music centre is a unique site in the dog's mind, probably as in humans in the most primitive part. This could lead to an explanation as to why 'gentling', and this often means playing classical music, can improve productivity in farm animals. Archbishop Isador's thirteen hundred-year-old dictum, 'Music soothes the mind to endure toil and the modulation of the voice consoles the weariness of the labour,' might just apply to dogs too.

VISION

In veterinary practice I am often asked to examine the eyes of what are perfectly healthy young dogs. The owners are concerned that their dog is going blind. In many breeds there is just cause for concern. Progressive retinal atrophy is an inherited blindness that strikes many dogs. The story I'm told is often the same. 'I was standing right in front of him but he didn't see me,' I'm told, or, 'I pointed to where his ball was lying but he sniffed all around and didn't seem to see it.'

In most cases the owner's concern is fortunately unjustified. Their pet dogs are simply manifesting their great difference in sight from us. Although the dog is perhaps ten times more sensitive to peripheral movement, he has poor vision up close, though reasonably good vision at a distance.

Most carnivores, dogs included, have more flattened eyes than humans have and although they can change the shape of their lenses and with that alter the focal length, they can't do it as well as we can. A dog's eyes are more sensitive to light and movement than ours are but their resolving power is correspondingly less. Technically speaking, sensitivity depends upon a large number of different types of receptor cells in the eye activating the optic nerve and the brain simultaneously, but no one type of receptor being able to do it itself. Visual acuity on the other hand depends on one specific type of receptor activating the optic nerve and brain. Our eyes are more acute than a dog's, which is why we can resolve better – and find tennis balls lying in front of our eyes.

The wolf has more lateral vision than many breeds of dog. The eyes are placed further apart but in many instances we have infantized breeds to have frontally placed eyes like humans. The dogs that were once called 'gazehounds', Borzois, Afghans and Salukis, all have frontally placed eyes. The guardian or minder dogs such as the German shepherd or the Akita have more laterally placed eyes while breeds such as the Cavalier King Charles Spaniel and the Boxer have eyes in an intermediate position. Many terriers have 'slanted' eyes. They are physically frontal but the slant allows them to see around corners.

The angle of vision varies tremendously from breed to breed, feeding varying amounts of different information to the dog's mind. If we take the visual angle of the human eye as our term of reference and it is 0 degrees, then the Pekinese have a visual angle of 5–10 degrees and terriers of 20–30 degrees.

The more that a dog sees laterally, the less well it sees straight ahead, and this is why so many dogs have difficulty finding something that is right under their noses. They have poor binocular vision, poor depth of field. Convergence, the focussing on an image in front of you, is most efficient in breeds like the poodle with their frontally placed eyes rather than in setters and retrievers with their more laterally placed eyes. The proportion of uncrossed fibres in the optic nerve is directly related to binocular vision and an ability to focus depth. In humans, almost half of the fibres in the optic nerve are uncrossed. In the cat, only a third are uncrossed and in the dog only one quarter.

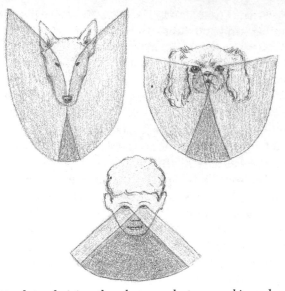

Dogs have better lateral vision than humans but poorer binocular vision.

The placement of the eyes will result in differing amounts of information being sent to the brain but curiously there is another effect. People frequently think of animals with frontally placed eyes as more intelligent and conversely those with laterally placed eyes as less intelligent. This attitude has an almost subliminal effect on the type of training we give our dogs and the responses we expect.

Dogs are exquisitely acute to the most imperceptible changes to their environment. They are masters at visually reading a territory or even an emotional state. A sheepdog's distance vision is so acute he can pick up his shepherd's hand signal at up to a kilometre distance. He can also see better than we can at night. Dogs are primarily diurnal rather than nocturnal which means that their eyes are equipped to work best in twilight. They do so because underneath the receptor cells of the retina, the rods and cones, is a glistening opaque reflective layer called the tapetum lucidum that reflects light back to the rods and cones. This is what causes the dog's eyes to shine in a car's headlights. The fact that a dog can also widely dilate his eyes allows as much light as possible to arrive at the retina. The receptor cells called rods, which register low levels of light but only in black and white, are necessary for night vision and dogs have lots of refractile rodlets in their eyes although nowhere near the fifteen layers in the cat's eye. This means that dogs cannot see when it's absolutely dark but are good at seeing in 'less light'.

The receptor cells called cones are necessary for seeing in light. They are also necessary for colour vision and by using sensitive training techniques, it has been shown that dogs can in fact see in rudimentary colour rather than simply in shades of grey. Experimental evidence shows that certain cones in the dog's eyes send messages to different cells in the lateral geniculate body, a way-station between the retina and the cerebral cortex. These cones are only sensitive to short-wave length light, or colour. The fact that the training in these experimental tests is exceptionally difficult means that colour vision is of little importance to the dog but they do, at least theoretically, have the capacity to see colour. (Most animals that hunt in daylight probably do.) However, they don't have a need for it. Biologically speaking, only animals that hunt by day and have a very varied diet need significant colour vision. This really means primates, monkeys and us. As we primates grow older, our lenses yellow and we lose violet in our colour spectrum which is why old painters rarely paint true violet hues. Dogs don't have to worry about such things.

My two retrievers have a varied diet, green grass, orange carrots, dark blackberries that they graze upon like cattle in the autumn but they still see this vivid world only in shades of grey. At least they can come home after a hard day in the park and settle down to watch show jumping horses or old re-runs of Lassie on television, something their American cousins cannot do. The final curious fact about a dog's vision is that their visual acuity is sufficient to resolve the European transmission frequency of 625 dots per second into a visual image. American television transmits at 525 dots per second, not quite fast enough for the canine eye to see anything other than a screen full of fast moving dots.

SCENT

Here are some facts:

1. The average dog has around 220 million scent receptors in his nose. We have around five million.
2. If the membranes lining the inside of the dog's nose were laid out flat, the total surface area of those membranes would be far greater than the total surface area of the dog's entire body.
3. The average dog has about seven square metres of nasal membrane. We have about half a metre and we're usually bigger.
4. The average dog has such acutely sensitive scenting ability that it can detect and identify smells that are so dilute that even the most sensitive of scientific instruments cannot measure them.
5. Scent is undoubtably the most important of the dog's 'practical' senses but also the most difficult for us to comprehend.

Odours play tunes in the dog's nose.

Odours have a powerful influence on both the physiology and the behaviour of the dog. It's no exaggeration to say that they are genuinely mind-bending. Smell memories last for life and affect almost all canine behaviours.

The dog's ability to smell the world around him and to interpret these smells depends upon a complicated chemical sensing system. First of all, he has mobile nostrils that help him determine the direction of the scent. Then he has the sniff, that marvellous disruption to the regular breathing pattern that in laboratory tests seems structured on a series of one to three successive trains of sniffing with three to seven sniffs per train. The most sensitive part of the dog's nose, the septal organ, is probably responsible for initiating sniffing behaviour.

Air that is sniffed passes over a bony structure called the subethmoidal shelf, a structure that humans don't have, and on to the lining of the nasal membranes. The area above the shelf is not 'washed out' when the dog expires air and this allows smell molecules to remain there and to accumulate. When a dog breathes in normally, the air goes through the nasal passages but continues down to the lungs. Sniffing leaves air 'resting' in the nasal chambers.

The numerous complex folds of the maxilloturbinate bones that provide the skeleton for the nasal membranes are designed to create airflow patterns that cause odours contained in air to strike regions of smell receptors. The odour molecules get dissolved and concentrated in nasal mucus which sticks to the receptor cells. Lots of mucus is needed for this. We produce about a pint of nasal mucus each day to help us smell things. Dogs produce proportionally far more.

Once the odour-laden mucus adheres to the fine microscopic hairs on the receptor cells, the chemical smell signal gets converted to an electrical

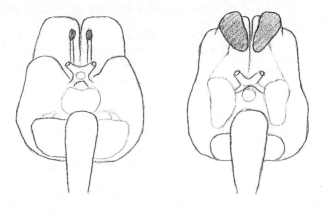

(i)   **(ii)**

Olfactory bulbs (inferior surface) (i) human brain (ii) dog brain. A much larger part of the dog's brain than ours is devoted to scent interpretation.

signal and is transferred to the cerebral cortex and to the limbic system, the part of the brain responsible for emotion. There are direct connections from the nose to the parts of the brain that regulate eating behaviour, drinking and sexual behaviour. Small odour molecules stimulate the olfactory nerve. Larger odour molecules stimulate the trigeminal nerve, one of the major cranial nerves, and volatile substances in sex hormones probably stimulate the vomeronasal organ, another piece of canine anatomy that doesn't exist in primates, but that I will discuss in more detail shortly.

Naturally, the receptor cells have to be exquisitely sensitive to detect the minute differences between a very large number of odours and this has become a fascinating area of study. As long ago as 1895, it was noted that dogs preferred odours of animal origin to other odours. The naturalist E. T. Seton in 1897 used a delightfully descriptive term for dog urine scent marking. He called the spots 'scent telephones'.

In 1908, scientists observed that dogs preferred natural to artificial musk and by the 1920s, controlled experiments were being carried out to determine exactly how sensitive a dog's nose was. In one experiment, a deodorized pine stick was held for two seconds in a subject's hand, was then handled by four other people and was finally placed among twenty other similar sticks. The dog was allowed to sniff the subject's body and then consistently found the handled stick.

H. M. Budgett's text, *Hunting by Scent*, is still a landmark in observations of the dog's ability to follow a scent. Budgett showed that dogs could follow a track even when the quarry put on rubber boots or jumped on a bicycle. He hypothesized that when one odour trail ran out, the dog started following another, the scent of broken grass for example. Budgett wrote that the best condition for a dog to follow a trail is when the ground temperature is a little higher than the air temperature – early evening. Not unexpectedly, that also happens to be the favourite hunting time for keen-scented carnivores.

By 1946, it was known that male dogs were better scent trackers than females, probably because they use their sniffers more every day in their territorial and sexual behaviour. More important, it was now also known that the dog's success at following a scent could be increased by training or by giving the dog small amounts of caffeine. In 1955, Kalmas Bjanbeh carried out experiments that showed that dogs recognize different odour secretions from one person's armpit, hand palm and sole of foot. He also showed that dogs could distinguish between odours of members of a single family but not between identical twins unless they were both present.

Research work was now getting more sophisticated. In one experiment carried out in 1961, dogs were allowed to smell progesterone, the female hormone that increases after ovulation. Once they had done that, they could detect articles that were simply handled by women in the post ovulation stage of their menstrual cycles. This was in fact scientific confirmation of the old herdsman's claim that some dogs could reliably pick out cows in estrous, cows about to ovulate, from the rest of the herd without being instructed as to which cows to choose.

In the 1970s, it was observed that dogs could detect butyric acid, a constituent of human perspiration at up to a million times lower concentration than we can and by the 1980s, it was known that the dog's ability to detect odours can increase tenfold if the dog is given these odours systemically. The enhanced ability was evident five to twelve days after it was administered and lasted for around seven days. This last research is the source for the unsubstantiated suggestion that drug detection dogs are actually given the drugs they are looking for, either by mouth or by injection, to enhance their performance.

One agency that has been avidly interested in the dog's ability to scent is the US Army. The US Army Mobility Equipment Research and Development Command at Fort Belvoir, Virginia, studied dogs and concluded that, 'The canine can be trained routinely to function extremely well as a Mine Detection System which is capable of operating in a vast expanse of climatic and topographical environments . . . (it is a) . . . highly adaptable, high sensitivity, high specific detection system which is relatively inexpensive, reasonably durable, readily reproducible and immediately available.'

The final report in 1985 from the Army Research and Development

Center said there was no mechanical peer in mine, booby trap, explosives detection. The Army incidently has tried other animals, badgers, coatimundis, coyotes, beagle/coyote crosses, deer, ferrets, red foxes, duroc pigs, javelinas, opossums, raccoons, spotted, striped and hog nosed skunks, all to no avail.

Being the US Army, they didn't test their dog's ability to scent by simply hopping on a bicycle as Budgett did. Instead they laid tracks across fields then deep ploughed the ground, covered it with petrol and set it alight. They sprinkled live and spent ammunition in minefields, had trackers use scuba apparatus and swim downstream underwater before emerging on to the bank on the opposite side but none of their ploys consistently defeated the dog's nose.

All of us are familiar with the air sampling strategy of dogs, how they sniff the air for plumes of odour, for clues on dust particles and water droplets. It's difficult for us to visualize these eddies of air because odour detection of this sort is foreign to our experience. One of the curiosities of canine scent is that male dogs make better trackers than female dogs. This contradicts the situation in many other species. We humans have a daily fluctuation in our scenting ability. It's at its best in the late morning and late afternoon although the morning high remains constant all day if we skip lunch. Women, however, can detect certain smells at lower concentrations than men, especially when their estrogen hormone level is high. More on that in the next chapter. Even more dramatically, female mice will alter their sexual cycles when they smell male mice and recently mated female mice fail to become pregnant if they smell strange males. The synchronization of estrus in dogs (or in secretaries in the typist pool for that matter) is almost undoubtably mediated by chemical signals, by scent, but as with almost everything else that affects the dog's mind the situation is not simply stimulus-response.

Urine from female dogs in estrous is only attractive to experienced males. They have to learn what the smell means. Urine markers and urine marking will mean nothing to a dog if he is raised in the absence of other dogs. They have to learn the significance of markers and learn how to mark. Response to scent certainly does begin early. Russian ethological studies report that dogs respond with salivation to meat odours at weaning even though they have no previous exposure to meat (although it's always possible that they have been conditioned to respond to similar chemicals in milk). Katherine Houpt's experiments on food preference in dogs revealed that they preferred pork over lamb or horsemeat, lamb over horsemeat, beef over horsemeat and beef over lamb and that these preferences were based on smell, not taste.

Larry Myers at the University of Alabama had developed electro-encephalographic (EEG) methods to evaluate the dog's ability to smell. He did so because, of all the dogs referred to his veterinary school with owner complaints about the dog's hunting or field trial performance, over 40 per cent were found to have an impaired sense of smell. Certain viruses such as parainfluenza can damage the nasal membranes and although nasal receptor cells replace themselves every two weeks, it takes about six weeks for a dog that has suffered a nasal infection to fully recover his scenting abilities. Myers states that cortisones such as prednisone might interfere with the dog's ability to scent as can many other medical conditions such as diabetes and epilepsy. To help owners determine the efficiency of their dog's scenting apparatus, Myers has produced a smell threshold test kit containing five different odours to home test a dog's scenting ability. Underlined and in bold type with the kit comes the admonition, 'The successful use of the kit requires that any dog tested be calm.'

Sensitivity to odours is partly inherited. In the grand genetic design of the dog there is already in place a magnificent scenting system, but even this can be improved upon through careful breeding and training. Bloodhounds and beagles are good examples of enhanced scent ability through breeding and the American Department of Agriculture's drug sniffing airport 'Beagle Brigade' is a fine example of the success of training. Odour is of unparalleled importance to dogs. Odours play tunes, as it were, in the dog's nose. They activate enzymes in the boundaries of the olfactory cells and elicit scent memories but they will do so more in one breed than another. Scott and Fuller, in their famous experiments in dog behaviour at Bar Harbour, Maine in the early 1960s, put a mouse in a one acre field with beagles. The beagles only took a minute to find it. Good scenters. Fox terriers took fifteen minutes to find the mouse and Scotties were never successful. One actually stepped on the mouse. Another reason why Scotties haven't made it as service dogs. Some breeds of dog such as the Bloodhound are best at ground scent. Others like the collie are superior at air scenting which is why British search and rescue teams prefer this breed. Air scenting dogs run along an air scent trail weaving back and forth, head held moderately high and circle when they lose the scent, using ever increasing circles until they pick it up again.

The largest question mark hanging over the dog's nose concerns its vomeronasal organ. This is a pouch lined with receptor cells and located above the roof of the dog's mouth and behind the incisors. It opens by ducts into both the mouth and the nose, has a thick blood supply and receives both myelinated and unmyelinated nerve fibres from the brain.

With all of that it must be doing something but nobody knows exactly what. In other animals, the goat for example, the vomeronasal organ is sex related. Goats exhibit an activity called flehmen in which they grimace by pulling their lips up, wrinkle their noses and raise their heads. It's thought that the grimace is used to bring odour into contact with the vomeronasal organ. Coyotes and jackals have been seen to flehm but dogs do not. Larry Myers feels that this organ seems to function primarily in the emotional behaviour of the dog through its reception of pheromones – body scents. The vomeronasal organ is not involved in the conscious appreciation of smells but rather, probably, in an unconscious perception of sex hormones. The information it receives is transmitted through the olfactory bulbs of the brain directly to the limbic system, the part of the brain that is intimately involved in the creation of a wide variety of emotional behaviours.

One interesting consequence of the dog's ability to use scent in his interpersonal relations with other dogs is research that is presently going on in France to determine whether dogs can scent different emotional states in humans. In his book *The Man Who Mistook His Wife for a Hat*, the neurophysician Oliver Sacks described the case of a man who could smell like a dog. The patient, a doctor himself, had stewed his mind with fashion drugs but when he was fit enough to return to ward duty, he felt that he had the ability to scent the emotional state of his patients. He said he could scent anxiety, contentment, sexual arousal. His enhanced scenting ability lasted a few weeks but as his brain repaired, he lost this capacity. Under the direction of Professor Hubert Montagner, a group of psychophysiologists in the south of France are looking at whether dogs can scent emotions in children.

Knowing that babies use scent to identify their mothers and that parents can identify their children by their smell, and knowing what the US Army amongst others knows about the dog's ability to scent, Montagner's researchers have set out to discover whether dogs can scent emotion – anger, sadness, joy, depression, mental illness, in people.

Most dog owners when asked will say that their dogs understand their moods and it's thought that they do so through the subtle visual signals we send, the sloped shoulder, the raised eyebrow. Dogs are, after all, inveterate and sophisticated people watchers. The French team's first experiments showed that dogs smell different parts of children's bodies under different behaviour conditions. They smelled the child's head after appeasing behaviour from the child but more often the upper limbs of the child when the child was being active, acting in an agonistic way. When there was no apparent behaviour from the child, the dogs would smell the trunk and lower limbs. Dogs rarely smelled the anogenital region of familiar children.

In their second experiment, published in 1989, dummies were dressed in the underclothes of familiar and unfamiliar children. Dogs explored the anogenital region of the 'unfamiliar' dummies but the head and trunk regions of the 'familiar' dummies. The authors concluded that the dog's mind is probably open to human social odours and that smelling these odours, 'enables the dog to decode and integrate human behaviours and perhaps emotional states and cognitive processes which underlie them.' If this is in fact the case, then the dog's sense of smell can be used in new and exciting ways in the future, not just for sniffing out mould in lumber yards in Sweden, or termites in Texas, or drugs at Spanish airports or injured fell walkers in the Lake District of Britain. The dog's nose might just possibly lead us to a better understanding of human emotion.

BODY RHYTHMS

Anyone with a dog knows how accurate his body clock can become. If six o'clock is feeding time and you haven't fed him by then, the well-programmed dog will be there at your side looking pleadingly into your eyes, and if he's really clever, he'll have his food bowl clutched in his mouth.

Dogs are creatures of habit but their habits are attuned to their biological clocks. Curiously, they either have similar biological clocks to us or have the ability to attune their internal clocks to our culture and lifestyle. Dogs can be trained to an accuracy of within one minute of a twenty four hour cycle.

The question of exactly what is a biological clock has fascinated science. The master clock is probably somewhere in the hypothalamus and acts through the pineal gland, synchronizing body rhythms through its hormone melotonin. Body rhythms can occur over varying lengths of time and the dog has the capacity to attune itself to cyclical events that occur anywhere from hourly to daily to monthly or longer. Kennel workers in quarantine kennels in Britain have observed that quarantined dogs will alter their behaviour, become more active or eat less, on the day of the routine visit of their owners. This might be weekly, two weekly or monthly.

The most common body rhythm is the twenty four hour one, called circadian rhythm. Circadian rhythm is why ill children are most likely to have higher fevers in the evening or why penicillin is most effective when given in the late afternoon, and very recently it was discovered that in at least one animal, the hamster, circadian rhythm is under the control of a single gene. It's a reasonable assumption that there is similar genetic control or circadian rhythm in the dog.

Research published in 1988 in the American journal *Science* dramatically showed how animal behaviour can be controlled by the genes. It has been known for some time that circadian rhythm in dogs is not controlled by external events. Although the obvious explanation for twenty-four hour cyclical behaviour in dogs would be the twenty-four hour day, dogs that were kept in constant conditions without any time clues, no darkness, no set feeding times, still, maintained measurable body rhythms.

The hamster is known to have a very accurate twenty-four hour circadian rhythm but in 1987, Martin Ralph at the University of Oregon discovered a hamster that appeared to have a twenty-two hour rhythm. This male hamster started using the running wheel in his cage two hours earlier each day. Ralph mated this male to a female hamster with a twenty-four hour body rhythm and to his amazement he was presented with a litter in which half had twenty-two hour body rhythms and half had twenty-four hour rhythms, exactly what you would expect if body rhythms were controlled by a single gene. Subsequent matings revealed that the 'faulty' gene was for a twenty hour body rhythm and that Ralph could accurately predict and produce hamsters with twenty, twenty-two and twenty-four hour body rhythms. He showed that in the hamster one gene controls the part of the hypothalamus that in turn controls the neuroendocrine system which is responsible for maintaining body rhythms. (Ralph almost couldn't publish his findings. In 1987, vandals broke into the animal care facility at the University of Oregon and released his carefully bred hamsters into the neighbouring forest. Fortunately Ralph had taken some of his prized animals home as pets.)

It is also almost undoubtedly true that like us, dogs have other fixed body rhythms. Through breeding, genetic manipulation, we have increased the frequency of estrus cycles in the female dog from the original yearly as in the wolf and Basenji to six monthly, proof that the dog possesses a six month cycle. And it's not just females that have this six month cycle. Women have thirty day menstrual rhythms but with these are associated changes in visual threshold, smell sensitivity, decreased hearing, taste sensitivities and preferences and lesser sensitivity to pain. Proof that not all of these are hormone related is the fact that thirty day cycles in pain sensitivity persist during pregnancy and are also found in post menapausal women and in men. As a similar cyclical curiosity, there is a higher rate of admissions to mental hospitals for both men and women around the time of each full moon. What this means from the point of view of understanding the dog's mind is that the dog can experience different effects from the same stimulus applied at different times during the body's natural cycles.

Free ranging dogs are most active in the morning, foraging and marking their territories. Pet dogs will show this behaviour too, simply by standing more during the morning than at other times of the day. Pet dogs will sleep about half of the day, will stand for about a quarter of it and spend the other quarter either sitting or lying down. About 80 per cent of sleep time is slow wave sleep which is the sleep of the mind. Rem sleep, rapid eye movement sleep, accounts for only about 20 per cent of a dog's sleeping time but this deep sleep is necessary for the replenishment of the neurotransmitters in the brain. A hormone, possibly serotonin, is the neurotransmitter that induces sleep and its requirements are constant – under body rhythm control. The French physiologist Henri Pieron was able to induce sleep in a dog by taking cerebrospinal fluid, the liquid transport of the brain, from one dog and administering it to another. More recently a peptide related to slow wave sleep, called Factor S, has been discovered excreted in urine.

The exact function of sleep is still a mystery but there is one intriguing theory, championed by Francis Crick the co-discoverer of DNA, that REM sleep, the sleep of dreams, occurs to help us to forget, to help rid the mind of excess information. The argument is that if a mammal doesn't have a mechanism for ridding its mind of excess information, then it would need a correspondingly larger brain to cope with all the retained information. Mammals that don't have REM sleep, animals such as the dolphin and the Australian spiny anteater certainly do have larger cerebral cortexes to their brains. Crick says that REM sleep is a kind of natural shock therapy. Dogs that are denied REM sleep certainly suffer from obvious behavioural disturbances and simply build up a backlog of REM sleep requirements. Their behaviour does not return to normal until the body rhythm's requirements for REM sleep have been made good.

There are many other unknowns concerning body rhythms. Mice, for example, are more active when the barometric pressure is rising. Horses are more active before storms and pigs are more likely to tail bite then too. Is it possible that dogs have an additional sense, an ESP?

ESP

The term ESP is emotive because it conjures up the unknown. What it really signifies is an animal sensation that we aren't clever enough yet to understand. For example, can dogs foretell earthquakes and if so, are they using ESP? Can a lost dog find its way home over hundreds of miles of territory it has never traversed before? How can the mind of a simple Monarch butterfly guide it from a milkweed growing in a field beside a summer cottage in Ontario, where it was born, to a hidden valley deep within the mountains of central Mexico. What is ESP?

The ethologist Nikko Tinbergen has answered this question as well as anyone. He said, 'If one applies the term ESP to perception by process not yet known to us, then extra sensory perception among living creatures may well occur widely. In fact the echolocation of bats, the function of the lateral line in fishes and the way electric fish find their prey are all based on processes which we do not know about – and which were thus 'extrasensory' in this sense only 25 years ago.'

About ten years ago a British veterinarian living in Spain wrote to the Royal College of Veterinary Surgeons Library in London asking them to do a literature search on whether dogs could foretell earthquakes. According to his letter, his dog Oliver, a four-year-old German shepherd, had been 'evidently aware of some abnormality in the environment some 48 hours before the event', the event being an earthquake. The library found references going back to 1880 in the journal The Veterinary Record.

Just before the 1880 Tokyo earthquake, cats ran about in houses trying to escape. The Veterinary Record correspondent concluded, 'There can be no doubt that animals know that something unusual and terrifying is taking place. The Japanese say that geese, pigs and dogs appear more sensitive in this respect.'

Another note in the Veterinary Record, quoted from The Times, concerned an earthquake in Chile. 'Before the shock in Chili (sic) all the dogs are said to have escaped the city of Talcahuano.'

R. N. Langridge, the British veterinarian in Spain, says that his dog Oliver wouldn't settle, was disquieted, was not himself and 'emitted sounds new to his vocabulary', so can dogs really foretell earthquakes? Do they possess a sixth sense? There is a long oral tradition of animal awareness of impending disaster and, after Chinese claims that animal behaviour reports had a role in the successful prediction of the Haicheng earthquake in 1975, scientific interest was stimulated in this phenomenon. One of the first consequences was Professor Helmut Tributsch's book, *When the Snakes Awake*, published in 1982. Tributsch, a professor of physical chemistry, dismissed ultrasonics, magnetic pulses or electrical disturbances as the stimulants of animal awareness of earthquakes. He believed that dogs and other animals are aware of electrostatic changes in the atmosphere that precede certain earthquakes and that this is their 'sixth sense'. He drew together anecdotes from different earthquakes before which animals had been reported to behave unusually. Before the massive earthquake in Tokyo in 1880, a giant horseshoe magnet temporarily lost its power. Before the Friuli earthquake in Italy in 1976, a local watchmaker noted that parts of a watch he was mending developed mutual repulsion, an electrostatic change. In 1988, at the Department of Physiological Sciences of the University of California, a baseline rate of randomly observed unusual animal behaviour was developed and using this guideline, Ben Hart, a veterinarian, carried out retrospective interviews to investigate animal behaviour prior to five moderate Californian earthquakes. Only one earthquake was preceded by a significant increase in the frequency of unusual animal behaviour and he concluded that this will only occur before certain types of earthquake.

The question still remains as to whether dogs have a sixth sense. There are innumerable stories of lost dogs finding their way home across continents, fording rivers, crossing busy highways, without a scent trail to follow, but successful nevertheless. It's been suggested that, like some birds, dogs can use the angle of the sun's rays, or like some sea water mammals, they use electromagnetic navigation to find their way across the unknown. The same question can of course be asked of us. How do Inuit people find their way home in a blizzard when there are no visual markers, when they are for all intents blind? The answer to both of these questions lies in the superlative use of the existing senses in conjunction with sensory capacities that we do not as yet fully understand. The dog, even the small spaniel on a satin cushion, remains biologically to his core an outdoor specialist. His mind functions according to the information it receives from his senses and as a domestic housepet living, in most cases, a sensory deprived existence, his capacities remain unfulfilled. The more sensory information that the dog's brain receives, the more developed his mind will become. Anatomically speaking, sensory stimulation causes nerve cells in the brain to actually grow and make new synaptic connections with other nerve cells. The network expands to accommodate and assimilate new information. This is why it is so important to provide a pup with a stimulating environment. Pups that are purchased and then left at home alone all day, with little sensory stimulation, will try to stimulate their senses themselves, often by being what we consider destructive. If they don't, they will grow up with smaller brains and restricted minds, poor examples of the potential of their species.

Chapter Four

Hormones and the Mind

Hormones have a profound influence on behaviour – and perhaps surprisingly, behaviour has a profound influence on hormones. They are mutually interdependent which is what makes the hormonal influence on the dog's mind so fascinating. Take for example the situation of the dominant male dog, the leader of the pack. (With pet dogs, the pack, regrettably, is often the human household in which he lives.) Outwardly it would appear that the dog has become dominant because he has a high level of the male sex hormone, testosterone. But all is not as it outwardly seems. A dog is not born with the genetic blueprint to have a high testosterone level. Rather, it is the circumstances of his existence, which teat he claims as his own, how his siblings and his mother treat him when he is young, how his human guardians treat and train him when he is older, all of these behavioural factors that affect the level of stress under which he lives. The dog that copes best with stress is the dog that becomes a natural leader, and the natural leader is then entitled to increase his testosterone level to help him maintain his position of authority. Technically, the phenomenon is called biofeedback.

As I have mentioned earlier, pups that are routinely exposed to sensory stimulation when they are young, that are exposed to noises, flashing lights and balance tests, grow up to be bigger animals with bigger brains and cope better when placed in new or unexpected situations. When confronted with a novel environment, these pups don't freeze or panic but instead they investigate and examine their new surroundings. They behave this way because of the hormonal feedback that is influenced by early experience.

Most hormonal activity in the body is under the direct control of the brain. The brain receives sensory information from the senses, passes this information into the hypothalamus which in turn is connected to the brain's hormone producing body, the pituitary gland. The pituitary gland produces hormones that in turn stimulate hormone production in

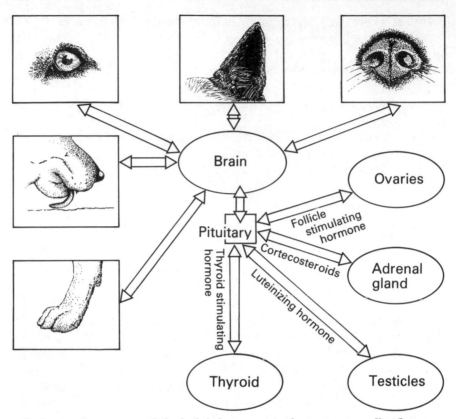

The brain, the senses and the body's hormone producing organs all influence each other through biofeedback.

various glands around the body. The pituitary hormone that signals the adrenal gland to produce cortisone is called adrenocorticotropic hormone (ACTH). ACTH released from the brain is related to increased excitability or anxiety in dogs while corticosteroids from the adrenal gland have a moderating influence on excitability. These two hormones normally control each other through biofeedback. When ACTH levels go up and excitability increases, corticosteroid levels go up and reduce the dog's excitability. The question still remains as to how early experience of sensory stimulation affects the dog's mind in such a way that he is later able to cope better with novel situations. The probability is that if pups are stimulated when their brains are still young, before they are myelinized, then the corticosteroid feedback to the brain will actually affect the development of the brain in such a way that ACTH production is finely tuned to corticosteroid stimulation. The system becomes delicately refined at a very early age resulting in more sensitive biofeedback.

Other pituitary hormones are also influenced by sensory stimulation and behaviour. A newborn pup's cry will stimulate the pituitary to release the hormone oxytocin which in turn acts upon the mammary tissue causing milk letdown. Growth hormone is released by the pituitary only after the pup falls asleep, perhaps one of the reasons why pups need to sleep more.

The pituitary gland is also responsible for producing hormones that stimulate the ovaries or testicles into hormonal activity. Wolves are sexually active only in the late winter and early spring, allowing for late spring litters to be born when the food supply is most plentiful. Increasing daylight stimulates a cocktail of hormones to act on the brain. A gonadotrophin releasing hormone trickles to the pituitary gland which in turn produces a follicle stimulating hormone (FSH). This hormone stimulates the ovaries to ovulate. With our domestic dogs, however, we have, through selective breeding, managed to override the seasonal light stimulation of FSH production in all breeds but the Basenji, and have replaced it with a six monthly biorhythm as its control (FSH production in dingos is still under light stimulation control.) This altered control over FSH stimulation is not always accurate which is why some female dogs will cycle every twelve months, the majority will cycle every six months and some will cycle every four months. FSH doesn't have a great effect on the dog's behaviour as such but rather influences behaviour through stimulation of hormone production in the ovaries. Similarly, the pituitary hormone, the luteinizing hormone (LH), has the same indirect effect on behaviour through its influence on testosterone production in the testes of the male dog.

Of all the hormones produced by the adrenal gland, those that are the most important from the viewpoint of dog behaviour are the corticosteroids. Dogs produce cortisol and corticosterone in their adrenal glands. These hormones have an anti-inflammatory effect and seem to diminish the allergic response. Because of that, corticosteroids are one of the most commonly used drugs in veterinary medicine, possibly accounting for more than half of all the prescriptions written to treat skin conditions in dogs.

Corticosteroids are released by the adrenal gland when a dog suffers a physical or a psychological stress. When the dropping apple frightened Lexington, her adrenal gland instantly pumped additional corticosteroids into her body. Now, if the apple had actually been a sabre-toothed tiger and Lexington had been savaged, then her corticosteroid production would have reduced the pain of her injuries, helped her to overcome and survive the physical stress while preparing her emotionally to fight or to flee. All of this would have instantly happened upon simply seeing the sabre-toothed tiger. The adrenal gland would immediately have pumped extra hormone into her circulation to help her fight or escape from danger.

A curious aspect of adrenal hormone production is its effect on sex hormone production in males. Studies have been carried out showing that tennis players, after winning a match, have an increase in testosterone levels while players who have lost a match have decreased testosterone levels. The same phenomena occur in monkeys (and politicians) and seem to be adaptive. If you're the winner, testosterone helps you to maintain your dominant position. If you're the loser, reduced testosterone levels inhibit a potentially losing challenge. This biofeedback mechanism is one of the reasons why it can be difficult to dislodge a dominant dog by being pleasant to him. The stress of defeat is often necessary to prepare him for his comedown.

Most of our pet dogs lead relatively simple and quiet lives in the luxurious dog houses that we call home. Nevertheless, the fight or flight survival mechanism of adrenal corticosteroid production still exists. But just as important, we administer massive amounts of corticosteroids to our pet dogs yearly without fully realizing the influence these drugs can have on their behaviour. The principle behavioural effect of these hormones is to reduce excitability in the dog and prepare it to stand its ground. In practice, this means that dogs with higher corticosteroid levels, higher either from their own production or through our administration, are more likely to behave aggressively. Owners sometime notice what they call an increased irritability in dogs that are on corticosteroid therapy. From a therapeutic viewpoint, corticosteroids are wonder drugs and are a prime therapeutic agent. They can, however, influence the dog's mind.

The fact that aggression is a commoner problem in male dogs than in female dogs suggests the importance of sex hormones on behaviour. On the surface it would seem that it isn't just hormones that cause aggression. Immature male pups are more aggressive and 'play-mount' more often than do female pups but the fact that castration is a common 'tool' used by animal husbandrymen throughout the world to control aggression in their stock is as good anecdotal proof as any of the influence of male sex hormone on behaviour.

Testosterone hormone production, as I've mentioned, is to some extent controlled by biofeedback involving the senses and other hormone producing glands. Something such as the scent of a bitch in estrus (or often the sight of the favourite stuffed toy or cushion), stimulates the hypothalamus to secrete releasing factors which in turn cause the pituitary gland to increase LH production and this in turn stimulates the testicles to produce more testosterone. Needless to say, this developmental pattern comes on stream at puberty which is why this time in the dog's life is one in which there can be dramatic behavioural changes. This is the time when, in his mind, he might choose to challenge his human pack members for leadership of the group.

Dogs need testosterone in order to develop normal male behaviour patterns but they also need to learn how to behave as males. An intact male dog that has never mated with a female can find it difficult, if not impossible, to differentiate between a female dog's head or vulva or his owner's arm. Male dogs that have been castrated and have no circulating testosterone, however, will continue to successfully mount and mate even though testosterone has virtually completely disappeared from the bloodstream within a day of castration.

The sex drive isn't coming from the adrenal hormones either. Dogs with their adrenal glands and their testicles removed will still continue to mount and mate if they have done so before surgery.

The effect of male hormone on behaviour probably begins before birth. Just before or around the time of birth the male pup is 'masculinized' by a surge of testosterone through his body. Female dogs do not undergo a similar 'femininizing' at birth, the phenomenon is restricted only to males. It's thought that this early surge of testosterone has an effect on the pup's developing brain, that it 'masculinizes' the development of the brain. There is also scientific evidence to show that nerve cells in the spine are affected by testosterone and that this is

The male pup's brain is 'masculinized' before birth by male hormone.

necessary if the male pup is to develop sexual reflexes such as mounting and pelvic thrusts. Strangely, it seems that testosterone, the male hormone, is actually converted to a hormone that is similar to the female hormone estrogen before it can influence the brain. Studies have shown that the administration of the female hormone estrogen can stimulate male sex behaviour such as mounting and even perhaps aggressive behaviour.

The fact that the male pup's brain is masculinized by male sex hormone before or at birth accounts for the common pup behaviour of clasping and mounting other pups or even people from as early as five weeks of age. It isn't 'instinct' that makes pups clasp and pelvic thrust. It's testosterone. Later on at six to nine months in small breeds, or nine to twelve months in larger breeds, when the pup reaches puberty, the remainder of his secondary sex characteristics will develop.

Even with the dramatic effects of testosterone on the neonatal dog's brain and the subsequent surge of sex hormone at puberty, male dogs that have not been adequately socialized to their own species when they were young will be poor breeders. They won't know what to do. And on the other hand, dogs that are frequently used for mating, stud dogs, will often mount any female regardless of whether she is in heat or not, simply as a conditioned reflex. As well, mounting is a dog's technique for showing dominance and can occur independent of sex. Dominant male or female dogs might mount others of their same sex not because they are homosexual but simply as a sign of pack seniority.

There are several false myths concerning sexual activity and emotional behaviour. Male dogs will NOT become emotionally more stable if they have sexual experience. That's in the mind of the owner, not the dog. Similarly, it's a fallacy that dogs should be castrated so that they will be less frustrated. That's also in the human mind not the canine one. Similarly, puberty does not signify emotional maturity. Some dogs can be mature long before puberty. Others will not mature until some time after.

Don McKeown at the Ontario Veterinary College lists seven facts concerning behavioural problems at mating.

1. A dominant male may not be able to breed an overtly submissive female (because she'll submissively roll over at the very sight of him).
2. The presence of a dominant male may inhibit a subordinate male from carrying out his mating duties.
3. A male dog may refuse to breed in a strange environment. (The upset of his mind can stimulate a surge of ACTH which can in turn inhibit testosterone production and mating behaviour.)

4. A stud dog may, for his own idiosyncratic reasons, show a preference for certain bitches and refuse to breed others (hence the heavy mascara that my younger bitch wears).
5. Anything frightening might inhibit sexual behaviour in the male dog.
6. Dogs are monogamous in the wild and even in 'captivity', females might show a preference for certain males and only breed with them.
7. In a similar vein, a dominant female might allow a subordinate male to mate with her.

Although the influence of testosterone on behaviour is well understood in many species of domestic animal, of all of them the dog is virtually the only one in which castration is not routinely carried out to modify behaviour patterns. It can be argued then that dogs don't suffer from a range of serious behavioural problems that are caused by the male sex hormone testosterone, although in practice I frequently see problems that can be reduced or eliminated through castration. More frequently though, I am asked by clients to castrate their dogs because of the mistaken belief that the surgery will correct the behavioural problem with which they have to contend, biting the postman for example.

The effect of castration will vary from dog to dog but the effect is certainly not a result of reduced testosterone alone. Circulating testosterone is gone in around six hours after castration but it might take six to twelve months before the castrated dog loses interest in sex. Others can continue mounting and mating normally and bitches in heat seem to be equally stimulated by castrated or entire dogs.

Ben Hart at the University of California carried out the most extensive surveys on the effects of castration on dogs and came up with these statistics:

BEHAVIOUR

Roaming
Reduced in 90 per cent of cases
Rapid reduction in 45 per cent
Gradual reduction in 45 per cent
No effect in 10 per cent

Intermale Aggression
Reduced in 60 per of cases
Rapid reduction in 25 per cent
Gradual reduction in 35 per cent
No effect in 40 per cent

Mounting People
Reduced in 60 per cent of cases
Rapid reduction in 30 per cent
Gradual reduction in 30 per cent
Some decline in mounting bitches in heat too

Urine Marking in the House	Reduced in 50 per cent of cases
	Rapid reduction in 20 per cent
	Gradual reduction in 30 per cent

Castration does not change a dog's personality, nor does it interfere with the abilities of working dogs whether they are service, hunting or guard dogs. Castration doesn't influence the dog's relationship with people either, except that it is ultimately more likely that the castrated dog will accept authority more readily from his human pack members than he would have before castration. Castrating before puberty will prevent a dog from fully developing his adult male physique and, as a general rule, dogs need a drop in calorie intake of about ten per cent after castration. This is due to an altered metabolism and reduced activity. The age of castration is not terribly important. The same effects will be seen at virtually all stages of the dog's life regardless of exactly when he is castrated.

Spaying female dogs, on the other hand, is a common procedure, carried out to prevent unwanted litters but equally to control the behavioural effects of the twice yearly surges in the female hormones estrogen and progesterone.

Estrogen is produced in the ovaries, stimulated into production by the pituitary hormone FSH. In all mammals studied, but with the exception of the human female, an increase in estrogen production is needed to allow a full display of sexual behaviour. Under the influence of estrogen, bitches might be more active, will urinate more, at least certainly in the presence of dogs, and will behave in a receptive manner with male dogs or even with their owners. A female in heat will sometimes cock her leg to urinate and seem to increase her territory marking with urine. Under the influence of estrogen, some bitches will whine and moan more, others will be more nervous and irritable, some will even mount males or non-estrous females. There are however certain false myths about breeding in bitches as there are with dogs. It is not necessary, for example, for a bitch to have a litter. It will not make her more emotionally mature or stable.

Under FSH influence, estrogen is seasonally produced in the ovaries of the dog but then, after ovulation, the other female hormone, progesterone, becomes dominant. Progesterone is fascinating. It probably has a calming effect on the dog's mind. Certainly in large doses it has a sedative effect and in fact, modified only very slightly, it is actually used as an anaesthetic in animals. While estrogen increases in the dog's body for a short length of time, progesterone remains in the circulation, influencing the brain for two months after each estrus and can have a dramatic effect on canine behaviour. The most common behaviours are those associated with pregnancy, nest building, guarding possessions and milk production. In most instances the behaviour is not associated with a true pregnancy but rather with a false hormonal pregnancy, a progesterone induced false pregnancy.

The behaviour of a bitch experiencing a false or phantom pregnancy can change dramatically. The sedative effect of progesterone on some bitches can be stultifying and it is not uncommon that pet owners bring in their post-estrous bitches thinking that they are seriously ill when in fact they are only subdued. This is a very subjective diagnosis. One of my bitches, Lexington, becomes overwhelmingly subdued for two months after each season, so much so that I am not afraid to call it a depression. She sits or lies under tables or behind furniture and flatly refuses to play with her senior partner, even if Liberty brings her a toy and drops it on her face. Then, as if by magic, over a period of a week to ten days, she snaps out of this state of mind and reverts back to her natural playful state.

Before I spayed Liberty, she too had behavioural consequences of her twice yearly surge of progesterone. Her taste buds changed. Starting immediately after the discharge part of her season, after she ovulated, she would start to separate her food, eating the meat and leaving the biscuit. If we substituted brown bread for the biscuit, she ate it ravenously. After this initial departure from her routine, she would then eat the bread but refuse the meat. If I changed brands and gave her a different tinned food, she was more than willing to eat it. I see similar problems in practice. Owners of bitches frequently telephone concerned that their dog has gone off her food. On asking whether she has recently had her season, the answer is often yes.

Guarding toys, dolls, rags, slippers or anything else that can be carried is another common behavioural consequence of the surge in progesterone. Progesterone, like estrogen, influences behaviour by acting on the brain. This is why it is now frequently used as a treatment in behaviour modification in male dogs. As I've mentioned, dog pups have a surge of testosterone around birth that 'masculinizes' their brains. Females don't have the equivalent. The lack of the secretion of ANY hormone at birth allows for the development of a 'female' nervous system.

Giving estrogen to a male dog can actually stimulate male sexual activity, but progesterone on the other hand seems to have a sedating effect and it is likely that this sedative effect makes the dog more amenable to behaviour modification therapy. The calming effect of the high level of progesterone is also the reason why it is best to avoid spaying bitches for two months after estrus. Spaying during this time can result in a precipitous drop in progesterone levels with possible accompanying emotional disturbances, irritability, aggression and depression.

Whether to spay, and if so when to spay, is a perennial question in veterinary medicine. If pet owners do not plan to breed from their bitches, I strongly recommend spaying in most instances before the dog's first season. Unlike male dogs, females come under no sex hormone influence until they reach puberty. Suddenly a short sharp surge of estrogen, then a prolonged two month surge of progesterone, dramatically alter the dog's behaviour until both hormones subside and the dog gets back to 'normal' again, back to 'anestrus'. A male dog lives with a constant supply of male hormone circulating in his body and influencing his mind. Females on the other hand only come under sex hormone influence twice yearly for a total of four months.

During that time, and specifically, under the influence of progesterone, there can be dramatic departures from previous behaviour patterns. The most unpleasant, often occurring in terriers, is the possessiveness that some can develop over objects. Once they develop this hormone induced behaviour, it becomes a learned behaviour that can continue for life, independent of whether the bitch is spayed in future or not. As long as a female is anatomically mature and doesn't have an infantile vulva, then the safest time to spay her is before her first season. That way she will probably never suffer from mammary tumours, the most common tumours in bitches, and she will never suffer the emotional upheavals of estrus and the consequent and normal phantom pregnancy. Spaying affects a dog's ability to work only in that it saves the dog from losing the ability to work because of hormonal surges. Spaying doesn't change the personality of a bitch. If anything, it preserves the natural personality of the animal. The only potential behavioural consequence of spaying is that inherently dominant bitches might become more dominant. Spaying prevents estrous cycles but it does not make a dog less feminine. It can't because in the absence of the testosterone influence at birth, a dog's brain simply IS feminine. Spaying can however stimulate a dog's appetite and alter her metabolic rate. This is why some spayed dogs become overweight. This potential problem can easily be avoided by reduing the calorie intake by around ten to twenty per cent after spaying and by ensuring that the bitch continues to get sufficient exercise.

Chapter Five

Communication

Dogs communicate with each other by using all the building blocks of their behaviour in an integrated fashion. Often through hormones as intermediaries, and by using the brain's integrated circuitry to its maximum they utilize a number of their senses simultaneously – sound, visible marks in their environment, facial expression, body position and scent. Of these, scent is the highest developed and probably the major means of communication, not surprising when you consider that the dog's sense of smell is the most highly refined sensory ability.

Facial expression, body posture and scent are all used in communication.

Non verbal communication through scent is mediated by chemicals that dogs produce called pheromones. We, as I've mentioned, produce pheromones too. They are responsible for the synchronization of menstrual cycles in offices where many women work in close proximity to each other, but dogs have a veritable cornucopia of pheromones that activate or inhibit other dog's minds. Pheromones are present in their saliva, faeces, urine, vaginal and preputial secretions, and in their anal, perianal and dorsal tail glands. They can influence immediate behaviour responses from other dogs or can initiate more long term responses.

Dogs' faeces act as calling cards for other dogs. They provide information of the name, rank and serial number variety to the sniffer in several ways. First of all, faeces get coated with mucus secreted by gland cells in the large intestine. Then anal sac secretions and products from the perianal glands are added. These pheromones give sex and social information. Sniffing the anal region is simply the canine equivalent of shaking hands. Just as we note whether a hand is sweaty, or a grip is firm, whether there is a wedding ring, whether a hand is soft or chapped, dogs derive similar information and distinguish individuals by their anal region smells. This is also why Fido your pet Great Dane might have what could be considered the antisocial habit of lifting up strangers by the crotch. It's also why retrievers prefer to carry in their mouths unwashed socks and underwear rather than cleaned articles. As Professor Montagner has observed, once the anal smells of the new dog (or person) are recognized, the sniffer goes on to sniff other pheromones from other parts of the body, the lips for example.

The dog's anal glands are his most important scenting apparatus. Anal glands in dogs are paired bulb-like reservoirs on each side of the anus. They are similar in position and to some extent function to the scent glands in skunks although they almost undoubtably emit a number of different pheromones. Writing in the Journal of the American Veterinary Medical Association C. A. Donovan described how an anal gland substance emitted by a dog under alarm in a veterinary office examining room was avoided by other dogs, but that the anal gland secretion from a female dog in estrus was sexually exciting to a male dog. The contents of the anal glands are not sex attractants but the physical character of the discharge changes with the sex cycle. Very little research has been carried out on canine anal gland secretions but the similar glands of the red fox contains at least twelve volatile components (the substances 'smelled' by the vomeronasal organ), most of which are saturated carboxylic acids and all of which have trimethylamine as their base. As dogs are so acutely sensitive to the tiniest chemical changes, it is probably true that anal sac secretion mirrors the dog's status and self-confidence as well as his sexual status. Small dogs that drag their bottoms along the grass or carpet, or larger dogs that obsessively lick their anal regions, are really trying to empty their anal glands. One of my retrievers has the curious and unpleasant habit of chasing her tail until she grabs it near the root and then pulling on it until her anal sacs discharge. Anal gland substance can vary in consistency from an almost clear colourless yellow, through a range of milky substances to lime green and on to brown or mahogany. The texture can be from watery to tar-like and all of these factors are a consequence of the rich bacterial flora and active fermentation that goes on within the sacs. A common surgical procedure in veterinary medicine is to remove anal glands that persistently block or get infected. Anal gland problems are more common in protected house dogs, perhaps because these dogs don't get frightened frequently enough for natural emptying. Problems also occur more frequently in non-breeding males and in spayed or castrated dogs although no one knows exactly why. However, knowing the importance they have in communication, it could be that a less invasive approach to medical problems involving the sacs such as antiseptic or antibiotic flushing might be more appropriate.

The function of perianal glands in dogs simply isn't known. These glands form an orbit around the anus and frequently develop hormone dependent tumours in older dogs. In 1988, it was discovered that the perianal glands in cows are a source of an estrous pheromone – just one more message for the bull to sniff. It's not unlikely that there is a related pheromone function in dogs.

The dorsal tail gland is a remnant scent gland in dogs. Older dogs, Labradors in particular, can lose their hair over this elliptical gland on their tail. In other canines, wolves, coyotes and foxes, the dorsal tail gland emits pheromone markers for trails.

Urine is the other major source of pheromones in dogs. Urine contains sex hormones and leaves information on the reproductive condition of the female and the power and authority of the male. Now if you're a dog and you want to leave packets of information concerning yourself for others to find, it's best that you leave them at nose level and this is why dogs cock their legs. In random studies of canine urine marking, dogs have been seen to urine mark eighty times in a period of four hours with the last markings absolutely dry because the supply of urine had run out. The frequency of urine marking is not constant however and is directly related to the early experience the dog has had in his life. Scott and Fuller's penned dogs at Bar Harbour didn't mark anywhere near as frequently as house dogs do. These dogs were penned in a 'clean' environment, an environment in which there were no other canine scent marks. Because they never experienced other dogs' scent marks, they never developed the habit of scent marking. These pups grew into adulthood retaining the squatting position for urinating. House pets on the other hand, especially urban housepets, frequently come in contact with other dogs' urine scent and will mark it. Ask any veterinarian what happens after one dog has cocked his leg in reception. The others form a silent queue behind him. It may well be that this type of marking is anxiety related. In fact, it's very possible that almost all types of urine marking, whether the cause might outwardly appear to be related to aggression or to sex, are simply stress reducing on the dog by surrounding him with familiar scent.

Females will sometimes cock their legs too but of course have a little more difficulty in directing their urine sideways. They don't scent mark as much as the males do, although some visit local scentposts near home. That's why females are easier to train to urinate in a specific spot. During estrus, females are more likely to wander and urine mark more frequently as a way of communicating their receptivity. This estrus urine is only attractive to sexually experienced dogs. Dogs with no mating experience will pay lots of attention to estrus urine but their minds don't integrate the pheromones in the urine with the receptivity of the female.

Males and females will sometimes mark their urine or faeces by scratching at the earth around them. These marks act like arrows pointing at the marker as well as spreading the scent around. It's also possible that another scent is left at the site from the sebaceous glands between their toes. The damage that scratching does to the nearby vegetation will also be evident to the sensitive noses of other passing canines.

You can't really say that dogs analyse the various pheromones they come across, but if the vomeronasal organ is involved in pheromone absorption as it probably is, then it's likely that the effect of these volatile substances works directly on the limbic system of the brain resulting in a specific emotional response, an unconscious response. Certain scents are obviously informative. The vaginal secretion methyl p-hydrooxybenzoate stimulates mating behaviour in dogs. But other scents, saliva, ear gland secretions, urine and anal pheromones probably act as hidden persuaders that influence behaviour by evoking mood shifts or, even more subliminally, by causing more long lasting physiological changes to the central nervous system or the endocrine system. To mate successfully, dogs must be psychologically as well as physiologically prepared for each other. This is why scent marking increases around estrus and why estrous bitches are attracted to male urine.

The importance of pheromones in canine communication becomes most evident when dogs unfortunately go blind. Long before their eyes first open, pups recognize their mothers through pheromones. Identity of self, other dogs, humans, nest, territory and sexual state of other canines can all be maintained by the blind dog through pheromone communication. This is perhaps the most important fact in the defence of the argument that the quality of life of a blind dog can still be good.

Although they unwittingly do so, females use pheromones deceptively in their communication with male dogs. They send scent messages fully ten days before they reach estrus and ovulate, telling the fellas that they are willing and able when in fact they are neither. If more than one dog is attracted, there can be competition between the males. Once the female is actually receptive to mating, she is more likely to mate with the strongest and fittest male. This type of chemical deception works in favour of the female but at the expense of the early bird male who gets replaced by a stronger male.

Pheromones influence the onset of puberty in males and females. And as I have mentioned, they influence ovulation and synchronization of estrous in female dogs. They can act as an aphrodisiac and by stimulating the sex hormones, augment the sex drive. They can also influence aggression in dogs. Pheromones communicate social status, emotional and physiological states, age and genetic relatedness. Their effect on the dog's mind is often unconscious but they are the most powerful form of communication that the dog has.

Voice, on the other hand, is probably the weakest form of canine communication but is still nevertheless much more intriguing that it might outwardly appear to be. A pup's crying, for example, physiologically affects his mother by stimulating her hearing in such a way that her brain passes information to her pituitary gland, triggering a release

Voice is the weakest form of canine communication but more intriguing than it outwardly appears to be.

of the hormone oxytocin, which in turn affects her mammary glands and permits milk letdown. All because of a certain type of vocalization, the cry.

Individual vocal differences in pups that allow the mother to differentiate between members of her litter probably exist at or shortly after birth. The mother's ability to recognize increases significantly over the next few days. All humans, but especially mothers, can differentiate between hunger, pain and pleasure noises from their infants. Experience helps in this matter but it's a safe assumption that all dogs and especially experienced mother dogs can make similar judgements on the noises their pups make.

Dogs make five basic sounds:

1. Infantile sounds – cry – whimper – whine
2. Warning sounds – bark – growl
3. Eliciting sounds – howl
4. Withdrawal sounds – yelp
5. Pleasure sounds – moan

As with scent, sounds can communicate individual, physiological and sexual information. Sophisticated analyses of wolf howls show that each howl is as unique as a fingerprint, that they are complicated sounds that probably serve several functions.

Communicating by sound has the obvious advantage of leaving the body free to do other things. A warning growl can be given as a dog positions himself to defend his territory. In the wild, sound communication had the advantage of not leaving a track for bigger predators, as scent communication does. Communication by sound is one canine trait that has been dramatically selected for in the domestication of the dog. Barking was encouraged through breeding and the dog's alarm bark was probably one of the first canine traits that our ancestors selected for. Dogs are far more vocal than their wolf ancestors although feral dogs are much less vocal than pet dogs, very Basenji-like in their economy of speech.

David Mech has analysed vocal communication in wolves and has observed that wolves will actually bark when strangers approach the pack. They will growl when there is a challenge for food and will whine when greeting each other or when their curiosity is stimulated. He says there are several different harmonic wolf howls, a loneliness howl, a 'pass on the alarm howl', a 'where are you' howl and a 'let's celebrate' howl.

Human intervention in dog breeding has accentuated all these forms of communication but none more than the infantile sounds of dogs. By neotenizing the dog, by perpetuating juvenile behaviour into adulthood, we have increased the frequency of crying and whining especially in certain breeds. What you will rarely ever see is an adult dog whining at another dog. They rarely whine at each other but rather direct it at us. This is really a learned response. Pups very quickly register that juvenile whines and whimpers are marvellous ways of getting our attention. This means that this form of communication can be controlled and diminished as long as we don't reward it with either interest, affection or food.

Barking can signify many emotional states in dogs. Its most obvious significance is as a warning, the alarm bark, but equally it can be used as an eliciting noise, a 'come on over to my house' call. Barks can signify threat, alarm, excitement or simply be attention getting and the knowledgeable dog owner soon learns to identify what different barks mean. Owners recognize their dog's territorial barks, their demarcate and defend noises but also their pet's ways of saying, 'Let me in', or 'There's a dog out there' or 'There's a human out there' or 'There's a salacious little bitch out there that I really want to meet.' Both of my retrievers emit deep, loud and sharp, snappy barks when they feel threatened or alarmed, but my older one, in the excitement of seeing another big dog, will solicit play from the stranger with a higher pitched bark, sometimes almost a yip. Growling however is an uncomplicated warning sound, often associated with warning facial and body expressions.

The howl is the classic eliciting communication in dogs and is common in wild canines and in some breeds such as huskies, malamutes, hounds and Dobermanns. Howling coordinates the pack members' spacing in their territory. Wolves howl to assemble the pack, pass on an alarm, locate each other, communicate personal information over vast distances, coordinate departures, reunions and movements and possibly to celebrate after the chase. Pack hounds certainly do this. A curiosity in dogs is their pitched howl in response to certain types of music. Singing canines sing by howling. It appears to be a pleasurable act to the singer but is usually thought of simply as a party trick by the howler's owner.

The most common withdrawal sound in dogs is the yelp. Dog owners need no training to understand that the yelp means either distress or actual pain.

Pleasure sounds in dogs seem to be unique to certain animals. The most common is the moan, a sound that sometimes mixes with an infantile high pitched whimper. As with infantile sounds, these too are learned behaviours, sounds only used on humans, never on other dogs.

The final way that dogs use their senses to communicate is visually. Dogs reveal their emotional state through the position of their ears,

Body and facial postures flow from one set of signals to another. This dog's mind is changing from calm and alert to aggressive.

This dog's mind is changing from calm and alert to submissive.

mouth, face, tail, hair, posture and body position. Emotions can USUALLY be accurately identified from facial and body signals but not always. That's one of the reasons why some breeds are more unreliable than others. Rottweilers, for example, are poor body signallers. They can move from contentment to anger without revealing the change in their body posture.

When dogs watch us, they watch our eyes. They do so because eye contact is an important means of communicating authority. The dominant dog stares down less dominant ones and the submissive dog avoids direct eye contact by averting his gaze and exposing his neck. Body posture signals flow through stages as the dog's emotions change. If his mind changes from fear to aggression or play to aggression or aggression to submission, his body and facial postures will flow from one set of signals to another. Dogs signal some of their feelings this way:

Calm – ears and tail relaxed
Alert – ears and tail up
Aggressive – hackles up, tail up, rump up, lips pulled back
Increased aggressive – snarl with teeth exposed, straight stance
Frightened – ears flattened back, tail between legs (This is also the fear
     biter posture if you approach past his critical distance)
Fear – crouched with tail between legs
Abject submission – lying down, hind leg lifted, urinates
Greet – lick face, beg regurgitated food or play bow

There is one final point concerning communication in dogs. As we know, they use their ears, tail, hair, eyes and other visual signals to communicate mood and emotion but think of how we have both genetically and surgically altered the dog's ability to speak through his body. Tails, the most eloquent of emotion messengers, are still amputated for the sake of vanity in many breeds. Others have breed standards that stipulate that their hair covers their eyes. The Old English Sheepdog is a classic example, so altered by our whims that it's sometimes hard to tell which end is which. Although if left to nature non verbal communication is a refined and sophisticated art in canines, our intervention in their breeding and in their morphology has dramatically influenced their abilities to send uncomplicated messages.

PART TWO

The Psychology of the
Dog's Mind

Chapter Six

Early Learning – Maternal and Peer Imprinting

The concept of 'critical periods' in the emotional development of the dog is a well documented one. Dogs that are denied human contact until they are over twelve weeks of age seldom make good companions. Somewhere in that time span is the 'critical period' during which dogs can be socialized to another species, us. Research into critical periods in the development of the dog's mind has been carried out since the early 1960s. In 1961, the magazine *Science* published the results of the most elaborate and definitive experiment that had been carried out to that date, a report that concluded that socialization in dogs, the ability to learn to live compatibly with dogs and with us ends at twelve weeks of age and that the most critical period was six to eight weeks of age.

Later on, in 1967, *Science* published again on the subject. The magazine reported Scott and Fuller's work which showed that pups raised in complete isolation to seven weeks of age could still recover completely and become socially normal. They also reported that outside contacts as infrequent as twice a week and for only twenty minutes each time were enough to ensure normal development as long as these outside contacts occurred in the critical period between four and twelve weeks. In the same year, other researchers reported that pups in this age range could form social attachments to another species, rabbits, simply by seeing them for as little as two hours.

Out of this and other research came the concept of the first critical period, which lasts from birth to twelve weeks of age, in the development of the dog's mind. It was divided up this way:

1. Neonatal period	0 to 2 weeks
2. Transitional period	2 to 4 weeks
3. Socialization period – to dogs	4 to 6 weeks
– to humans	4 to 12 weeks

This is an excellent concept and a good canvas on which to draw an outline of the development of canine learning and behaviour. Its most significant flaw is that it states in bold numerical figures what happens and when. This is fine as a concept but all too easy to misinterpret as fact. What we must remember when looking at the early development of the dog's mind is that each pup has had its own unique prenatal environment. Each pup is already under the influence of his own genes as well as his mother's hormones. Some have a considerable headstart before they even emerge into the world. This is why we should first look at the prenatal influences on the dog's mind.

THE PRENATAL PERIOD

Most research on the effect of the prenatal environment on the subsequent development of the mind of the newborn has been carried out on rats. In this species there is some evidence that mothers that are highly stressed during pregnancy produce young that subsequently have lower breeding success. There is also evidence that in certain lines of rats, if the mother is stressed during pregnancy, her rat pups grow up as more fearful animals. Other research shows that if the mother is stressed during the third term of her pregnancy, her pups will show reduced learning ability, extremes of behaviour and increased emotional states.

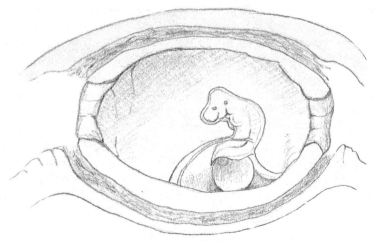

A pup's mind is already being modified by the uterine environment.

Although most of the evidence comes from research with rats, it's only logical to assume that environmental factors that affect the pregnant bitch can also affect the subsequent development of her pups' minds. This is why proper nutrition, good exercise and warmth are so important during pregnancy. This is also why it's important to avoid unnecessary x-rays, drugs and chemicals and to prevent parasites and disease.

THE NEONATAL PERIOD

Pups are born with immature, still developing brains. Although their brains look like jelly, they're really more like sponges soaking up what they can from the environment. During the neonatal period, that's not very much, simply because their sensory abilities, hearing, seeing and to some extent feeling, smelling and touching are not yet well developed. The pup is almost wholly under the care of the mother and the way she behaves with her pups will influence their behaviour in later life.

The Swedish Dog Training Centre in Solleftea, Sweden is responsible for training police, guide and search dogs in that country and for some time now has had a breeding programme producing 300 to 400 German shepherd dogs each year. At Solleftea, they discovered that certain behaviours in pups, whining for example, were maternally imprinted, not genetically predetermined. They discovered this by synchronizing births in German shepherd bitches and then immediately exchanging some pups from each litter. By doing so, they observed that the maternal influence on the subsequent development of the pup's mind was far greater than had ever been expected.

Konrad Lorenz developed the theory of imprinting in 1937 when he observed that young goslings raised in an incubator in the absence of their mother followed him around if he was the first living animal they saw. He was 'imprinted' in their minds as their mother and for the rest of their lives, they identified with people rather than with other geese. Once they matured, Lorenz had to contend with sexual advances from his more ardent feathered admirers.

In the context of maternal behaviour in dogs, Lorenz has hypothesized that the short forehead of pups, the cheeks thickened with fat sucking pads and the erratic movements of the neonate pup are all factors that elicit maternal behaviour from their mother, that imprint the pups on her mind (and on ours for that matter because we respond to the same stimuli too).

But just as the pups imprint themselves on their mothers, so too at the same time the mother is imprinting behaviours in the plastic and formative minds of her young. Remember, these first weeks of the pup's life are its most important, that the earliest experiences it has will have a tremendous impact on the mind of the pup.

Maternal behaviour is more than simply hormones. The pituitary hormone prolactin can induce maternal behaviour in bitches but it won't induce maternal behaviour in male dogs, even male dogs that have been castrated at birth. (These dogs are already masculinized by the prenatal surge of male hormones.)

There is no apparent maternal behaviour centre in the brain and it seems that both genes and learning are involved. Monkeys, for example, have to have been mothered themselves in order to 'spontaneously' become good mothers and it seems that the same applies to dogs. Smell is also involved. Maternal behaviour can be induced in virgin rats or even in males by the simple presence of newborn rat pups. This suggests that the higher portions of the central nervous system are involved. In wolves, males are normally involved in the 'maternal' behaviour of regurgitating food for their pups. Nursemaid 'aunts' in the wolf pack do the same thing and care for pack pups when mother is searching for food.

Maternal behaviour can't be hormonally induced but equally, the bitch with her first time litter doesn't need any experience to know what to do. There are, however, variations in the quality of mothering and these will affect the behaviour of her pups. What differentiates maternal behaviour in dogs from wolves is the fact that for thousands of years we have intervened in the process, from choosing the nesting site to building the nest, helping at birth, cleaning, feeding and grooming the young to culling the litter. The result is that there are quite a few potential mothers around that actively need our help in order to mother successfully. In selectively breeding for certain behaviours or body shapes, we have unintentionally impaired normal mothering ability, especially in some breeds. I will discuss breed differences in Chapter Ten.

Normal maternal behaviour involves nest building 24–48 hours before birth, a restlessness for about 24 hours and then finally parturition. The mother licks the pups dry but otherwise ignores them until all have been born. This licking is important for the mother as it imprints the pups into her mind. This is why there are more rejection problems for pups born by Caesarian section where the mother can't carry out this imprinting procedure. The mother cleans up her nest by eating the afterbirths – good housekeeping. Sometimes mothers become excessively tidy, eating the afterbirths, chewing up the umbilical cords but then eating the pups. This form of cannibalism occurs most frequently in bull terrier breeds and is probably a genetic fault. Other forms of cannibalism, horrible as it may seem, are normal.

Hamsters and mice use cannibalism as a standard method of litter size control. It's possible that bitches do the same. It can be argued that it's even beneficial to the mother and the survivors for her to do so. Reducing the number of mouths to feed and feeding herself at the same time seems to be a rather good survival tactic. Bitches certainly rid the nest of the sickliest by moving them away from the others and leaving them to die. This is normal sensible canine behaviour for the experienced mother but can also occur erroneously with immature or emotional bitches that isolate healthy pups this way. In wild canines, infanticide, the killing of a litter of young by another female, does occur but this practice is seldom, if ever, seen in domestic dogs.

Sometimes bitches will move a litter just after birth to a safer nest. They don't carry pups by the scruff of the neck like cats do but rather grab any part of the pup's body and carry or drag it to the new site. Several years ago, I attended to a Cairn terrier who, perhaps because of her displeasure at having so many strangers look in at her newly delivered litter (she bore her litter at the residence of the American ambassador), deftly transported them all down a rather magnificent cascading flight of stairs, across a ballroom floor and out into the garden where she secreted them, and herself, deep in an earthy rabbit warren.

Once the litter is born, the mother licks and grooms the pups, stimulating them to urinate and defecate but also imprinting on their developing minds their first outside message, 'I'm your mother and this is what my saliva smells like.'

This chemical clue is vital for the pups' survival. Mothers lay a saliva trail by licking the pups and licking her own nipples. If her nipples are cleaned with soap and water, the pups can't find them. But if they're washed in the mother's own saliva, the pups have no difficulty. This is why it can be so frustratingly unsuccessful to try to get a pup to suckle by simply sticking its mouth on a teat. It doesn't have the smell stimulation.

Sucking is not done simply for nourishment alone. It's also a bonding mechanism that identifies the mother to the pup. Animals fully fed by stomach tube still suck as much as animals that have not been fed. The pup treads and kneeds on the teat to stimulate milk letdown (which is why all dogs but especially bitches like to have their bellies rubbed). Pups incidently don't get attached to certain nipples but when they are older, dominant pups might express their superiority by denying a certain teat to others in the litter.

Lactating bitches are 'buffered' against dramatic emotional responses while they are in the active feeding phase of motherhood. With rats, more than one pup is necessary to stimulate the mother rat to act maternally and to secrete milk. This may well be true with some bitches too. Under the right stimulation however, throughout the two week neonatal period, the bitch wakes up her pups through licking and general cleaning, stimulating them to suckle. Her hormonal feedback system suppresses her normal adrenal gland response, diminishing her 'fight or flight' behaviour in the presence of danger. Later on, this will create the conflict that will ultimately lead to her weaning the pups.

Because the pup's brain is still in such a formative state, the experiences it undergoes now will have a tremendous impact on the development of his mind. This is why the first twelve weeks of age are such a 'critical period' in the pup's life. By controlling his environment we can actually influence the final form and structure of his mind. As I've mentioned previously, early handling – stressing – at this age is actually good for the emotional development of the dog and probably makes him better able to cope with stresses later on in life. Mild stresses early in life influence the adrenal-pituitary system, fine tuning it to respond in a sensitive and graded manner later on in life rather than in an all or nothing fashion. According to electroencephalograph (EEG) readings on pups, dogs mature faster if they undergo mild stresses early in life. They also perform better at problem solving when they are older than do other dogs. The question remains as to exactly what is the optimum amount of stress that a young pup should receive because it is also known that too much stress at an early age leads to subsequent retarded development. Michael Fox, for example, stimulated pups with stroking, clicks, flashes of light and balance. He found that these pups subsequently developed larger adrenal glands and when they grew up, were invariably dominant in competitive situations. In 'detour' problems later on in life, these handled pups 'kept their cool' better, made fewer errors and problem solved faster and quicker.

The evidence is firm that mild stresses even in the neonatal period are good for the development of the pup's mind, that constant temperature, comfort and freedom from adverse conditions does NOT make better pups. Mild stresses will accelerate body growth, reduce emotionality and possibly increase resistence to certain diseases. (Rats are less likely to get conflict situation induced gastric ulcers.) The question remains, exactly how much stress is good?

The early life of a pup is normally littered with human contact. We intervene in the life of newborn pups almost as much as we do with our own children. This 'incidental' handling, unnatural in the sense that it only occurs in dogs reared by us, is probably on average enough in itself to beneficially stimulate the development of the pup's mind. What is most important to understand about the neonatal period of canine development is that a hands off policy, a policy of 'leaving it to nature', of allowing the dog mother to behave 'naturally' is definitely not in the interest of the pup. The pup's mind will develop faster and further through sensitive early handling. We can help by providing a good environment but more than anything else, the pup needs the right mother.

During the neonatal period, the mother is the overwhelming influence on the pup's development. The appealing and apparently random movements of the pup are in fact part of a delicate pattern that imprints the pup in her mind. At this time the pup is still cut off from stimulation from the environment simply because his brain is so immature and his senses are underdeveloped. His survival is completely dependent on her. Even after feeding him, because his gastrointestinal tract is so poorly developed, his mother must lick his ano-genital region otherwise he can't urinate or defecate. She continues consuming all of her pup's evacuated waste for some time, stimulating his body functions and keeping her nest clean at the same time. (This normal behaviour can sometimes occur in non nursing mothers. We naturally think it's disgusting but only because it's not part of the human repertoire of natural behaviours.)

During the neonatal period, virtually all social interaction is with the mother. The other litter mates and we are of little importance. The mother is responsible for warmth, food giving, grooming, contact comfort and retrieving. His brain is soft and jelly like, barely myelinated and there is almost no EEG activity. Although he's sensitive to touch, to pain, to temperature and to taste, his hearing, vision and temperature regulation are still undeveloped. All of this, however, rapidly changes during the next two weeks of the pup's life, the transitional period.

THE TRANSITIONAL PERIOD

This is when most of the pup's sensory abilities come on stream. The eyelids open and the optic nerve becomes myelinated. The ears open and teeth appear. The EEG shows that the brain is stimulated by the sights and sounds of the world around him. In a twinkling, the pup's world opens up and suddenly his littermates and the rest of his environment have a dramatic effect on his developing mind.

During the transition period, pups start to wag their tails. They growl and bark for the first time. They notice us. Their temperature regulation mechanism is improved and this means they can start to leave the nest and will now, if possible, eliminate outside of the nest.

Perhaps the most dramatic changes in the pup's perception of the world around him are demonstrated on EEGs. During the neonatal period, the pup has an irregular low amplitude EEG with few alterations when the pup is aroused. The only sleep they experience is deep sleep, the activated sleep of dreams. As the pup reaches two weeks of age, however, the EEG shows that they go through a quiet sleep phase before falling into deep sleep.

VISION

Functioning of the pup's visual system can be monitored by an equivalent of the EEG called an electroretinogram (ERG). An ERG can first be recorded in the pup at ten days of age and by fifteen days, it has the basic features of the adult pattern ERG. By 28 days, the ERG pattern is fully developed. Translated into action, what this means is that at about two weeks of age, when the eyes actually open, the pup's response to light and to moving objects is rather weak and variable, but by four weeks it is similar to that of the adult. It takes a little while longer, however, another week, before the pup readily and easily recognizes from a distance his mother or owner.

HEARING AND BALANCE

The equivalent method of monitoring hearing in pups uses the brainstem auditory evoked response (BAER) test. Although loud noises can stimulate a response from the neonatal pup, it's not until the young animal is twelve to fourteen days old that his external ear canals open and that a brisk startle response to a loud noise will be evident. It's not until this age too that the pup shows any BAER activity, but quite suddenly by three to four weeks of age, BAER activity has all the characteristics of older animals. Similarly, by this age the pup has an adult-like vestibular reflex, righting and orienting himself in an adult manner.

TOUCH AND PAIN

When pups are born, only their heads have touch reflexes. But by two weeks of age, their front legs have touch reflexes and by three weeks, so do the hind limbs. This is why pups can sit at two weeks old and stand at three weeks. It only takes a few days more for their minds to coordinate all four legs and get them walking.

Pain sensation is present at birth but very slow. By the transition period, however, the pup's pain response is similar to the adult, fully matured and by four weeks of age, most of the brain is myelinated and ready for complex learning.

Maternal influence on the development of the pup's mind still remains strong but during the transition state, the influence of the litter mates and of us starts to increase. During the neonate stage, the mother stimulated her pups to nurse but now, in the transition stage, the pups start nursing on their own. Soon mother will be evading the little suckers and rightly so. This is when mother-young conflict first appears.

From the viewpoint of the young, they would like to suckle as long as possible. From the mother's viewpoint, however, she should wean them as soon as possible. Aside from sapping her of energy, we know that suckling also has a biofeedback effect that inhibits her adrenal gland's 'fight or flight' response. So it's doubly in her interest to wean her pups as soon as she can. From the cost/benefit view, it is during the transitional period that the mother beings to opt for independence from her pups. There are ways to trick this biological programming however. If old pups are taken away from a mother and substituted with younger ones, she will continue lactating longer to properly feed these youngsters.

Regurgitating food for pups is a natural behaviour we have selectively bred down.

During the transition period, some dogs will start regurgitating food for their pups, a wolf trait that we have probably intentionally bred out of most breeds of dogs because we find it unpleasant. If the pups don't eat this food, the mother will. This is the basis for the common behaviour that many dogs have of eating their own vomit. It is also during the transition phase that pups will start to beg food from their mothers. Wolf pups do this by pawing and biting at the lips of adults returning to the den. By selectively breeding dogs for thousands of years for tractability and obedience, we have unwittingly perpetuated many juvenile characteristics in their behaviour. As I have previously mentioned, wolves and related canines are naturally neotenized animals – the behaviour of the young, playfulness and curiosity are perpetuated in adulthood. In dogs this is even more so. The dog that jumps up to 'kiss me hello', as so many of my clients tell me, is really asking you to regurgitate a meal for it.

Dogs that jump up to 'kiss' their masters are really asking them to regurgitate a meal.

The transition period from two to four weeks of age is the beginning of the most important period in the young dog's life. For the first time his senses are being stimulated and he will be forever influenced by the images that now form in his mind. Experiments with rats dramatically show that animals raised in a sensory rich environment, in the case of rats with toys, tunnels and runways, develop thicker cerebral cortexes, have more synaptic contacts between neurons and have higher levels of neuroendocrine transmitters in their brains than do rats that have been raised in a bland environment. If nothing else, this shows just how spurious is the argument between instinctive and acquired behaviour. How a dog behaves at any given time in his life is a result of a constant and fluid interplay between his genetic potential and his environment. To the semanticist, the only things that are actually coded in a dog's genes are various amino acid sequences in protein molecules. These in turn influence the development and function of the dog's brain, his senses and his hormonal system and all of these in turn affect his behaviour. But, as we have seen, at every stage of the development of the dog's mind, he receives a feedback from his environment and this can profoundly affect the course of his development and his unique character. It is during the transition stage that many of the building blocks of the dog's future behaviour are laid down. What he experiences now will affect him for the rest of his life. How that actually comes about is still a deep mystery. I grew up in Canada. Into my mind is locked the fragrance and the feel, the sight and the crunch of autumn leaves. Not a single molecule that was involved in laying down that experience still exists in my brain but the experience remains forever. It's the same with dogs. No one knows exactly how it happens but as the pup leaves the transition period, he enters what will be the most important eight weeks of his life, the time during which he learns to live with both his own kind and with us.

THE SOCIALIZATION PERIOD

The social life of the dog can only begin when he has developed all of his communication facilities. As we have seen, most of these have matured to near adult conformation by the time the pup is four weeks old. His senses have matured – he can see, smell, hear and touch more than adequately. His brain is well myelinized so he can process information quickly and efficiently. And although the major hormonal influence on the dog's mind will have to wait until puberty, the male pup has already had a surge of male hormone course through his body, sensitizing him to behave in a masculine way. This is why, during the socialization period, male pups will mount other pups.

Any type of social relationship that a pup eventually develops begins as a problem and ends as a habit. These relationships are always adaptable and it is during the next two months that they are most malleable. Until now his only really important relationship has been one of care-dependency with his mother. During the socialization period, this relationship will gradually evolve to one of dominance-submission. According to Erik Wilsson at The Swedish Dog Training Centre it happens this way.

Wilsson studied social interaction between mothers and litters of German Shepherd pups from three to eight weeks of age. At the beginning of this period, while the pups were still in the transitional stage of development, the relationship was strictly care giving by the mother and care seeking from the youngsters. Behaviourists call care giving behaviour 'epimeletic' and care seeking behaviour 'et-epimeletic' but I'll use the simpler terms.

Maternal care giving is influenced by the considerable number of variables we now understand, genetic, hormonal, sensory and environmental. It consists of:

1. Licking the pup's ano-genital region and consuming their waste products
2. Licking their faces and generally grooming their bodies
3. By pushing the pups with her nose, positioning them for both warmth and feeding
4. Carrying those that have strayed back to the nest
5. Guarding her pups
6. Lying on her side to allow the pups to suckle
7. Vomiting food for her pups
8. Carrying food for her pups.

Care seeking behaviour from the pups is also a result of genetic and sensory influences but is much simpler and initially consists of:

1. Whining
2. 'Rooting' around the mother – crawling forward with the neck extended searching for warmth and a teat.

In the socialization period, care seeking behaviour becomes more sophisticated and consists of:

3. Tail wagging – with the tail held low
4. Yelping
5. Licking the mother's face, nose and lips
6. Jumping up
7. Pawing
8. Following the mother like glue

The obvious beginning of the change in relationship from care-dependency to dominance-submission occurs when the mother starts to walk away from the pups as they try to nurse. This usually occurs during the fourth to fifth weeks. Wilsson has observed that the amount of nursing, the time for final weaning and the evolving interaction between the mother and her pups during this period can have a permanent effect on the minds of the pups. He looked at six different maternal behaviours, 'nursing time', number of 'inhibited bites', 'growls', 'mouth threats', 'nibbling' and 'licking' of puppies.

Wilsson noted that when pups tried to suckle, some bitches 'punished' their pups with 'inhibitcd bites' more than others did. These pups started to show passive submission by lying on their backs to be licked earlier than other pups. 'Growls' and 'mouth threats' also lead to passive submission. Wilsson studied 600 pups and observed that the incidence of 'inhibited bites' and 'mouth threats' from the mother reached their maximum by the time the pups were seven weeks old. The care dependency relationship had evolved to one of dominance-submission.

The severity that the mother uses in altering this relationship has a direct bearing on how that pup will ultimately behave with people. Some of Wilsson's 'mothers' were extremely aggressive with their youngsters and continued to punish them even after they withdrew from her. Others were more benign, showing less aggression and vigorously grooming the pups afterwards. Wilsson says that there were stronger social bonds in these litters and that these mothers were more likely to 'paw' their pups into submission. He concluded that pups from litters that had been subjected to a lot of 'inhibited bites' were less socially gregarious with people than other pups – that they were less likely to approach a passive person. In 'fetch' tests, fetching a tennis ball, these pups were less likely to perform than others.

The way a mother alters the care dependency relationship to one of dominance and submission has a life-long effect on the dog's mind but as with everything else, there are no cut and dried answers. Learning about dominance submission through maternal punishment is absolutely necessary for the pup. It was discovered long ago that pups raised under non-punishing conditions were later impossible to train. Wilsson's studies which show that pups with aggressive mothers are less likely to 'fetch' demonstrate how inhibiting too much aggressive dominance from the mother can be.

Some people feel that the concept of 'dominance' and 'submission' is too restricting to use to describe the evolving relationship between mother and pup. Roger Abrantes in Denmark says that what a pup is really learning when his mother snarls at him as he tries to suckle is not fear but compromise. Abrantes calls this 'subdominance' rather than submission, a concept that I rather like. He says that relationships are not as simple as dominant and submissive, but rather are multilayered. This is a sensible hypothesis and it explains how there can be a pecking order within a pack of dogs or within a pack of people and dogs. In veterinary practice I often meet pet owing families who can lucidly describe the pecking order in the family and exactly where their dogs fit into it. Sometimes a dog might actually be at or near the top. In other circumstances it can be in the middle of the totem pole.

The dog is unique in the animal world because of our intervention in his breeding. Through breeding for tractability and obedience, we have unwittingly selectively bred an animal that has the potential for a life-long dependency upon others. Neotenized animals have a life-long need for us. We take on the 'maternal' responsibilities for feeding, grooming and housing our pet dogs. This need is tenuous at best. Dingos and Asian Pariah dogs are classic examples of breeds of dog that have successfully reverted to independence. The same is true of the feral dogs that now breed successfully in some regions of Central and South America. Whether or not these dogs will fetch is irrelevant. Their role in the pack is of most importance and their relationship with their lit-termates during the socialization period is the cornerstone for this. How a pup treats and is treated by his littermates during the next few weeks of his life is just as important for us too. We are, after all, his surrogate pack. To understand the dog's mind we need to understand the compli-cated relationships that evolve between littermates during their formative socialization period.

Play stimulates communal behaviour while it moulds physical and mental dexterity.

The litter is a complex but temporary society and in it play activity is the most common and important interactive behaviour. Playful activity has a number of functions. Socially speaking there are three:

1. Play stimulates communal behaviour. It creates social bonds with other dogs, bonds that would be necessary for the future well-being of the pack. The fact that there never will be a future pack, just us, is irrelevant to the dog's mind.
2. Play affects and moulds adult social behaviour. It's through the trial and error of play that pups learn their communication skills.
3. Play predicts the future dominance relationships within the pack. Playful activity is a good indicator of which dogs are more dominant and which are more submissive.

Play of course has many more functions, all of which have an influence on the dog's behaviour. Konrad Lorenz says that playful animals are 'specialists in non-specialization'. Through play they learn to manipulate inanimate and social objects. They learn to manipulate US!

4. Play promotes physical dexterity and mental flexibility
5. Play improves coordination
6. Play permits experimentation under safe conditions. It teaches the dog how to time himself, how to intercept or intervene, how to maintain his balance.
7. Play teaches action patterns. Dogs learn to carry out sequences of events.
8. Play allows for safe exploration.
9. Play stimulates inventiveness. It teaches problem solving.
10. Play is a lifelong activity in dogs. Adults will often play when 'invited' to do so by a younger dog and adults will play for no apparent reason other than the apparent joy of playing.

Play is a natural inherited canine activity, as strong in wolves as it is in Yorkshire terriers. The naturalist Erik Zimen once watched five young wolves at a lake in British Columbia play almost without interruption for five hours, chasing back and forth, jaw wrestling and playing 'king of the castle' on a large rock in the water. This is learning for life. Running, chasing and ambushing will be necessary for running down deer, caribou and moose when they are adults.

The wolf is a naturally neotenized species. Adult wolves play but we have exaggerated this behaviour in our pet dogs. When dogs play, the standard rules of interrelationships are temporarily suspended. My dominant dog Liberty intentionally falls to the ground and bares her neck in a self-handicapping manner for the subdominant dog Lexington to chew on. When they play, they have a lightness of being. They dance through the air. Reality is fleetingly suspended but comes back instantly if play gets too rough or if they are distracted. In classic canine play, normal behaviours, sexual, predatory and aggressive, occur but out of context. Michael Fox says that animals that play more, that have more sensory stimulation during the critical socialization period, grow up to be 'more intelligent and highly evolved'. Play develops the pup's mind because it leads him into different situations where he has to innovate solutions. It accelerates experience.

Some canine play seems to have become modified simply as an end in itself. Pups learn the play bow during the socialization period, the invitation to either mother or a littermate to join the activity. In adulthood, one or the other of my dogs chooses when to play and approaches the other. An approach can be made with a toy in the mouth, to play tug of war or simply 'you can't have it'. Use of an object in social play allows the dogs to play more vigorously. After all, objects can be pulled, chewed, carried and thrown better than partners can. Alternatively the approach might simply be to tumble, roll and jaw wrestle. Their playful activity is carried out with just a little more exuberance than would be necessary if it were for real and their roles pass back and forth. One dog will intentionally release her prize ring, ball or rubber toy so that the other can pick it up and will then try to retrieve it.

Some naturalists feel that wolves use play antics to 'fascinate' rabbits towards them. The wolf acts in a frivolous way and, by its bizarre behaviour, reduces the escape distance of the rabbit. This is possibly the basis of behaviour for the unusual Canadian hunting dog, the Nova Scotia duck tolling retriever. These dogs sit in their blinds with their masters until ducks land on nearby salt marshes. They then leave their blinds and act like fools, 'tolling' as their masters call it, barking, jumping up and down, running in tight circles. The birds take flight, are shot and then retrieved by the now saner acting canines.

During the socialization period, play is carried out for more than its own sake. It is a source of skill, information about littermates and it deflects natural aggression. It can also be used to advance the social status of a pup. Adversaries can be tested in a ritualized manner through play fighting. It teaches cooperative behaviour. Although there are no sex differences in play activity during the socialization period, it is still necessary for the development of normal sexual and social be-haviour. Dogs that are isolated from other dogs during this period in their lives are often hyperaggressive towards other dogs. They haven't learned how to inhibit this behaviour. Harry Harlow's unpleasant but informative experiments with monkeys showed that without early social experience, monkeys became withdrawn, introverted asocial and even autistic. They became poor mothers too. Social activity somehow also teaches animals how to be good mothers. Pups that are denied play activity up to twelve weeks of age can develop bizarre behaviours including self-mutilation in order to reduce tension. They are poorer learners, have a greater fear of people, animals and noises, are shyer and more antisocial. They will avoid stimuli and are reluctant to explore. In short they are canine misfits.

Play activity during the socialization period establishes social relationships and teaches the 'inhibited bite' – the soft mouth, as well as greeting rituals. Pups that do not play with other pups at this stage can become excessively or abnormally attached to humans and can be fearful of other dogs. This is one reason to avoid orphan pups or runts. The opposite occurs too. If pups are kennelled throughout their socialization period, it is more likely that they will develop a general fearfulness of strange environments, be excessively excitable or excessively inhibited. They will be poor learners and will try to avoid stimulation. Between three and eight weeks of age, pups should be exposed to potentially fearful stimuli in the environment, kids, aerosol sprays, vacuum cleaners, vets, postmen, cats, street noises, and this sensitization should continue throughout the socialization period up to twelve weeks of age and on into the juvenile period. Remember, this is the most sensitive period of a pup's life. Dogs that don't meet people until after the socialization period are antisocial, difficult to train and dingo like in their fight, flight and freeze behaviour. Dogs that don't meet other dogs during the socialization period are fearful, make poor mothers and are inhibited or over reactive when they meet other dogs.

From a veterinary viewpoint, a major conflict arises from this information for, as all dog owners know, standard medical advice dictates that pups should not meet other pups until at least two weeks after the final puppy vaccination which is given, at the earliest, at twelve weeks of age. Some vaccine manufacturers suggest isolation until a final parvovirus inoculation is given, as late as twenty weeks of age. But what happens to the pup's developing mind if he is isolated for so long?

The answer to this conundrum depends on what role you want your dog to play in your life. For dogs that will be forever human oriented, this medical advice can be safely followed. The consequence will be a human oriented and human attached dog, one that should be responsive to command and readily obedient to you but possibly fearful or aggressive with other dogs. He might also prefer your arm or leg as a sex object rather than a nice warm fellow canine. If you want your pet to be a dog oriented dog however, the standard medical advice is dangerous to follow. Both of my dogs were introduced to dozens of canines, healthy canines, during their socialization periods and the consequence is that they actively search out and enjoy the company of other dogs when they are exercised in public parks. This is doubly important if your dog is to be worked with other dogs. Avoid other dogs if the incidence of infectious disease is high in your area. In most countries of northern Europe and in many areas of North America, infectious disease incidence is low enough to allow your dog to socialize properly during this formative stage in his life.

The degree or intensity of play at this stage is also a portent of the future. Pups that are too playfully aggressive during the socialization stage make difficult pets in the future. This is when they learn most about dominance and submission. If there is any one rule to follow at this stage of a pup's life, it is a simple one. Never, ever play fight, especially during the socialization period, with potentially aggressive and dominant dogs. If you do you will be creating dramatic problems for the future.

DOMINANCE AND SUBMISSION

Dogs are gregariously sociable animals. Their minds must be plastic enough to permit the development of a dominance hierarchy amongst themselves yet, at the same time, allow them to seek each other out for comfort. My dogs know that my wife and I and our teenage children are all 'pack leaders', but the dogs would still like to sleep in our beds with any of us if we let them. In order to establish this pecking order of dominance and submission, dogs have a sophisticated system of mostly mute signalling and communication that becomes well developed during the socialization period.

Take 'threat' for example. The young pup can display 'threat' in a number of ways. First of all, he can display his weapons – show his teeth. He can display his maximum size through piloerection, raising his hackles. He can startle his adversary with a growl or a bark. He can feign an attack. All of these behaviours in pups look comical to us but they aren't comical to the littermates. This is dead serious business. During the socialization stage, a body language display of dominance and aggression can actually lead to an outright attack, a genuine display of aggression to establish dominance. The dominant pup will initiate an activity, stare at and perhaps stand over his littermate, then growl and possibly attack.

During this stage a dog 'learns' to signal dominance but equally he also learns to signal the opposite, friendliness, lack of aggression, submission, tolerance and cooperative behaviour. Dominant activities, what the behaviourist calls agonistic behaviour take the following forms. All of these are apparent during the pup's socialization stage.

1. Stalking other pups with head and tail down, hindquarters raised and ears erect
2. Chasing, ambushing or pouncing upon littermates
3. Standing over the littermate, head and tail erect, neck arched. The dominant pup often forms a 'T' shape with its adversary by placing its head over the other's neck
4. Circling the littermate while stiffly wagging the tail and walking imperiously
5. Actual attacking and biting, especially around the neck and face
6. Hackles up – piloerection
7. Baring of weapons – snarling and showing incisors and canines
8. Direct stare at the littermate with dilated pupil – 'showing eye'
9. Shoulder slams or less frequently hip slams
10. Standing with forepaws on the littermate's back
11. Boxing
12. Mounting either with or without pelvic thrusts
13. Wagging only tip of erect tail
14. Ears completely erect OR completely flattened against head
15. Play fighting taken to excess

Using these displays, the dominant dog implies 'power' to his littermates and we do exactly the same later in the pup's life when we start training him. By learning the importance of these body signals now during the socialization state, the likelihood of later fighting is dramatically diminished. After all, the pup's hormonal feedback system is still being developed. Puberty is still months away. The pup that copes best now will have the best developed adrenal-pituitary system and after puberty, this in turn will allow him to produce more male hormone which will reinforce his dominance.

There are many different types of aggression in dogs and I will discuss these individually in Chapter Eight. There is, however, a relationship between aggression and eating and this can reach a critical point during the socialization period. The hypothalamus is the part of the brain that when stimulated provokes eating. Greater stimulation to the same area, however, doesn't provoke greater hunger. It provokes aggression. Predatory attacks can be initiated by stimulating the hypothalamus. On the other hand, stimulating the ventromedial part of the hypothalamus inhibits aggression. Some research in cats has suggested that the chemical serotonin, a natural brain neuroendocrine, inhibits aggression in that species. A consequence of that research has been an examination of what kinds of diet may affect brain serotonin levels. This is still a completely hypothetical field but it does raise the interesting possibility that the dog's developing mind can be altered by its diet as well as the other known factors.

But just as important as the dominance behaviours are the defence and submission behaviours. These surrendering behaviours are vital to prevent unnecessary injuries or even death, and they too become finely developed during this two month period up to twelve weeks of age.

Pups show submission in these ways:

1. Tail between legs, ears depressed, head hung low, eyes averted from dominant dog
2. Submissive grin with lips retracted back
3. Licking lips, sometimes sneezing and showing incisors at the same time
4. Rolling on to back
5. Lying on side, lifting hind leg and exposing genital region
6. Urinating
7. Defecating
8. Remaining stationary while aggressor circles or places paws on shoulders
9. Remaining stationary while aggressor mounts

Dogs make such good companions because they are pack animals. They thrive on togetherness and this too has its formative development during the critical socialization period to twelve weeks of age. During these weeks, when the pup's mind is most impressionable, he learns to participate in group activities with his mother and his littermates. Behaviourists call these allelomimetic activities. They are often combined with other exploratory, social, care giving and care receiving activities and include:

1. Sleeping together
2. Feeding together
3. Walking, running, sitting or lying together
4. Investigating together
5. Barking or howling together
6. Grooming each other
7. Sniffing, nosing, pawing or licking each other

All of these group coordinated behaviours develop during the socialization period as normal litter or pack activities and are then either perpetuated by the dogs or co-opted by us when we become members of the 'pack'. That's why pups want to sleep with their new masters and why it can be beneficial in training to let the newly acquired pup actually sleep in the same room as his new owners. It cements the new forming bond between them. I did this with my dogs, using 'crate training' to condition their toilet habits at the same time as emphasizing that I was the new pack master, the 'alpha' animal. The new pup was given a very large cardboard box to sleep in. Half of its floor was covered with newspaper and the other half with a blanket. Because pups will do almost anything to avoid soiling their bedding, each one immediately took to urinating and defecating on the paper and sleeping on the blanket. And because they were in the same room as other members of the pack, because they could hear us and smell us, there was virtually no 'separation anxiety' noise after the first night. Later on, when they were old enough to control their bladders and bowels overnight they were given baskets to sleep in and moved out of the bedroom. (And later still, because their baskets took up so much room in a small house, they were moved back in, but that's a logistical and behavioural problem on my part, not on my dogs'.)

Feeding behaviours are also set in the dog's mind during the socialization period. Because they are pack animals, dogs are competitive feeders. In the litter they compete for the best teats. At weaning, they learn to beg for food from their mothers, a behaviour based on the wolf's routine of regurgitating food for the young. These activities are rewarded during the socialization period, are imprinted in the dog's mind and perpetuated for life. The consequence is that many dogs 'wolf' down their meal as if it might be stolen if they aren't speedy enough. Others become supreme beggars, experts at the soulful, mournful look that melts the heart of the malleable master. Pups from large litters, and that often means from larger breeds, have a greater tendency to gorge themselves on food. This might be a result of more competition for food during the socialization period. It is also probably genetic in origin. Pups from breeds that are more inately aggressive also bolt down their food faster than others of the same size. West Highland terriers, for example, are good gorgers. They are also highly competitive within the litter. Cavalier King Charles spaniels, on the other hand, are both less aggressive and more selective about their food. This behaviour is naturally modified to a great extent by the different types of owners that choose and then feed these breeds, but part of the difference can still be traced to competitiveness during the socialization phase.

Physical activities like walking or running together have their genesis during the socialization period and these too are adapted by us to use in the training of pups. To train a properly socialized pup to 'Come', all you need to do is take it to an open space, put it down, then walk or run ahead, turn around, wave your arms around and call it. Because it's been properly socialized to participate in group activities, it will automatically come to you. But if it has been isolated from other dogs and from people, its response will be impaired. Isolation for as little as a week during the socialization period results in impaired learning ability. The dog's mind is so affected that changes are actually seen on his EEG. The amount of human contact that a pup needs during this stage of his life varies from breed to breed but there are simple rules to follow. No exposure to humans during the socialization period will lead to a life-long fear of us and will make them difficult or impossible to train. Never isolate pups during the socialization period. Never leave them alone for extended periods of time. Don't use isolation as a form of punishment (Not yet, that is. Later in life short term 'symbolic' isolation can be a classic form of punishment.)

Pups can become bonded to almost any species during early socialization to twelve weeks of age.

Properly socialized pups enjoy the company of others, both canine and human. Pups that are socialized to other species at this formative time in their young lives become bonded to them too. My previous retriever had a companion rabbit that she used to play with and sleep with. Dogs can bond to cats and often do. Cats make excellent companions for dogs if the introduction take place during the early part of the socialization period.

The ability of a working dog to perform well has its origins during the socialization period. Group investigation is the basis for activity amongst many different types of service dog. To set a dog's mind to work with others, he must have the right experiences during this early stage of his life. The American Guide Dogs for the Blind programme carried out a detailed analysis of what factors make a good guide dog and came to a profound conclusion. They were concerned because a proportion of otherwise perfect dogs failed their final tests, because they were unable to take independent decisions, decisions such as refusing to obey a command that might endanger a blind owner. Their research narrowed down this lack of confidence to a single factor. Almost 90 per cent of pups that left their puppy kennels either at or before twelve weeks of age graduated as successful guide dogs. But only 30 per cent of the pups that stayed in their kennels for three or more weeks longer succeeded. A difference of only three weeks at this time in their lives was the major contributing cause of later training failure. Paced juvenile experience during the critical socialization period is of utmost importance for companion as well as service dogs for, just as the failure rate for guide dogs dramatically increased if they didn't socialize properly to people and the environment, so too will companion dogs be less trainable if they are kennelled for too long.

This is a fact that has been oberved by animal behaviourists on both sides of the Atlantic. Dr. Victoria Voith, the most respected canine behaviourist in the United States, and Dr. Roger Mugford, the most prominent behaviourist in the United Kingdom, both say that it is critically important that pups are obtained at around seven to eight weeks of age, during the peak of the socialization period. Fortunately, many pedigree pups are purchased at this age. The tradition ensures the greatest likelihood that the pup will grow up normal, stable and well adjusted. Both Voith and Mugford have written, however, that the most likely candidates that they see because of behavioural problems are cross breed dogs from pounds and dog homes. As it seems highly unlikely that there is a genetic basis for their observations, it is most probable that these dogs have had either interrupted or unfortunate socialization periods, leading to their later disruptive or uncontrollable behaviour.

It's doubtful that there is any age after which a dog is not trainable. Even adult wolves have, with patience and care, been socialized to humans. The end of the highly critical and most important socialization period, the period that on average lasts from four to twelve weeks of age, appears to be at least partly a consequence of what happens during that period itself. The important point to remember with the concept of critical periods is not the time intervals themselves but rather the concept that there are major developmental periods in a young dog's life, periods during which his formative mind, that composite of senses, hormones, experience and basic building blocks, is most susceptible to influence. The timing of these critical periods will vary across breeds and between individuals. It will vary according to the experiences that the pup has during these periods. Between three and five weeks of age, a pup will make a positive approach to anything or anybody that approaches him, but after about five weeks the fear response will develop. When the pup is still very young his recovery from the fear response will be almost instantaneous but later it takes longer.

From a veterinary viewpoint this is an important consideration because pups need their first inoculations against distemper and parvovirus at just about the worst time when it comes to fear imprinting, eight to ten weeks of age. Pain or fright at this age can result in a life-long fear of the veterinary clinic. Inoculations are necessary evils and my personal way around the problem is to use as thin a needle as possible, room temperature vaccine and most important, a distraction, usually in the form of a yeast flavoured vitamin and mineral tablet. Scott and Fuller found that it was easy to addict their dogs into the habit of eating sardines. I get pups into the habit of eating yeast tablets and while they are eating them, I give the inoculation. The ruse usually works and the pup subsequently remembers the food rather than the injection.

Young pups will only make contact with complete strangers for a very short period in their lives. This is a perfectly normal adaptive and protective behaviour pattern, allowing and ensuring early socialization with his own species but protecting him later on from potential predators. We take advantage of this natural behaviour mechanism by implanting ourselves during this critical stage in our dog's minds. By doing so, however, we become part of their world and are expected to conform to it. A dog doesn't expect to be treated like a human. A dog expects a human to act like a dog, to participate in group activities, to play, to hunt together, to sleep in the same den. The most essential feature of the dog's new environment, however, is us. Dogs, being the magnificent observers that they are, respond to the most subtle of changes in their owner's status, in fact just as children respond to changes in the emotional states of their parents. There is now convincing evidence that certain owner attitudes are directly associated with certain behaviour problems in dogs. The dog owner would generally prefer not to believe that his state of mind has a direct bearing on his dog's state of mind, that he can be the cause of his dog's misbehaviour. Similarly, dog breeders are loathe to accept that there are any bad dogs, only bad dog owners. Both attitudes are narrow and myopic. Genetic factors certainly do endow the individual dog with certain tendencies and predisposition to behave in particular ways. Maternal, environmental and peer pressures also alter the dog's behaviour but ultimately, and this is where dogs differ from most other species, ultimately the capacities and tendencies of the dog and the full development of his mind depend on the kind of relationship he develops with his owner. This develops more fully in the next critical stage, the juvenile period from twelve weeks to sexual maturity but before we look at that there is one question that still needs answering.

Our understanding of critical periods in the development of the dog's mind, and specifically our understanding of the socialization period, has led to attempts to develop methods of selecting ideal pups on the basis of their behaviour during the middle of their socialization period. After all, we know that dominance, subdominance and submission behaviours among pups usually develop between four and six weeks of age, that human attachments develop best between roughly six and eight weeks of age and that fear imprinting most often occurs first between eight and ten weeks of age. We know that pack behaviour has developed by five weeks and by seven weeks the group will be attacking the weaklings of the litter. We know that pups become attached to both their living and non-living environment by about seven weeks of age. Scott and Fuller gave us the concept of critical periods. The work of Guide Dogs for the Blind built on that, showing how important it is to intervene at the right time in a pup's life. Michael Fox's work on canine behaviour further popularized the notion of predictability in puppy behaviour and by the 1970s, 'puppy tests' became a popular and frequently recommended technique for selecting a suitable companion. I used a puppy aptitude test (PAT test), developed by William Campbell, when I selected Liberty, my older Golden Retriever.

Puppy aptitude tests are carried out on pups at around eight weeks of age, out of sight and sound of the mother, littermates, other dogs or other distractions. Each pup is taken individually to the test area and handled gently at all times. I tested Liberty myself. This is what is recommended. The tester should be a stranger to the pup. Campbell's five test components are:

1. Social Attraction
The tester claps his hands gently.
2. Following
The tester walks away from the pup.
3. Restraint
The pup is rolled on his back and held in that position for thirty seconds.

4. Social Dominance
The tester gently strokes the pup from the top of his head, down his back towards his tail for thirty seconds.
5. Elevation Dominance
The pup is cradled under his belly and held just off the ground for thirty seconds.

I carried out these tests and graded the litter of pups using Campbell's methods. Gradings range from 'highly dominant' through 'very dominant, dominant, submissive' to 'very submissive'. Liberty scored 'submissive'. She lay flaccid on her back when I pinned her shoulders to the ground, didn't squirm when cradled above the ground and came to a hand clap and followed me about. She liked being stroked.

When Liberty matured, however, she was anything but submissive. She became a dominant dog, not with us but with other dogs and with strangers. She became territorial and a guarder, a marvellous housedog. Her personality was so different to what had been revealed by her puppy aptitude test that we temporarily changed her name from Liberty to Bert.

Just how accurate then are puppy aptitude tests? Margaret Young at the veterinary school at North Carolina State University has evaluated the predictability of these tests. She carried out aptitude tests on several hundred pups at six to eight weeks of age and then again when they were three years old. On over one hundred of these dogs she carried out intermediate testing at sixteen weeks, twenty six weeks and one and a half years of age. Her conclusions were striking. 'Social attraction, following and acceptance of stroking,' she writes, 'did not reliably distinguish puppies that were later aloof and independent from those that were attracted toward people, readily trainable and handleable. Nor did tendencies identified by the tests at seven weeks as dominant or submissive reliably predict later tendencies toward dominance or submissiveness.'

Puppy tests didn't predict dominance aggression either but then again that's not surprising. Dominance aggressive patterns, as I will discuss in Chapter Eight, don't usually occur until the dog is emotionally mature and that is usually between one and three years of age. Dominance aggression is more readily predicted by size, sex and hormonal status and is frequently directed at members of the household as well as other dogs. Males usually dominate females. Dogs of like size, age and sex are more likely to fight because of unclear demarcation as to who is the dominant one. An interesting observation in 1987 from the veterinarian Leo Lieberman was that a survey of 200 owners of dogs that were neutered when they were six to twelve weeks old reported less aggressive behaviour (and fewer overweight and medical problems) than in dogs neutered at six months of age.

One aspect of puppy testing was a useful predictor of future behaviour. Few pups in Young's tests showed any signs of aggressiveness such as growling or barking at the tester. Those that did during testing continued to do so later in life. Of all the dogs tested, these were the ones that ultimately needed the most expert handling.

On the other hand, some pups that showed no signs of aggression during testing turned out to be real troublemakers when they matured. Possessive aggression, for example, over toys or food began appearing at four to six months of age, long after purchase. Protective aggression was seen in young pups from guarding or herding breeds but even here only occurred after six months of age. Similarly, fear induced aggression, punishment induced aggression and predatory aggression all developed long after puppy testing had been carried out. On the basis of these results, puppy testing is only of value if you actually come across aggressive or dominant behaviour in a pup, a rare event.

The dog's mind, it seems, remains plastic and malleable long after the socialization period has finished. In fact, many of his behaviours develop later in life once he is emotionally and sexually mature. Human attachment, however, is easiest at this stage and can be facilitated if the pup suffers an emotional stress during the socialization period, a stress requiring a greater than normal dependence upon us. Loneliness, fear, or injuries that are healed with affection, contact and attention all facilitate a stronger bond between dog and master. Pups that have the misfortune to need intensive medical attention when they are very young often grow up as animals with a greater dependency upon humans. Care can be the glue that bonds the young pup to his master for life. The seeds are sown in the socialization period from four to twelve weeks of age. The mind develops to maturity during the next critical period to puberty.

Chapter Seven

Later Learning – Our Influence on the Developing Mind

Although the patterns of his future behaviour are set in the dog's mind during the socialization period, it is over the next months that he learns the relevance of these behaviours and how to use them. His learning capacity is great and the speed with which he learns is considerable. Learning is fast because the blackboard, his mind, is still relatively uncluttered. The more he experiences life, however, the more difficult it becomes to instil new thoughts and ideas in his mind.

The greatest problem in understanding the dog's mind now occurs. It's our problem, not the dog's, for it is we who make the mistake of treating dogs as not quite as animal as other animals. The American psychiatrist Aaron Katcher described our quandary as lucidly as anyone when he wrote that pet owners look upon their pet dogs as 'four legged Peter Pans caught between nature and culture'. Dogs aren't caught between nature and culture. They might seem almost human to many of us but that's only because we recognize human attributes in their behaviour, fear, hunger, the desire to care for others and to be cared for in return. These aspects of their behaviour are all part of their culture and, although we recognize them, they might be displayed for slightly different reasons than those we have.

The unfortunate consequence is that many people genuinely do, during this formative period of the dog's life, treat him as a member of the family, democratically issuing conditional clauses to their pets when they misbehave. 'If you jump up on the dining room table once more Fido, I'll be really angry!' As hard as it might be to emotionally accept the fact, we must remember that the dog's thinking process, cognition, is not as complex as ours. Dogs don't think symbolically. To them fear is fear, sex is sex. What you see is what you get. The only exception to that is the occasional manifestation of psychosomatic illness, for example a feigned lameness to get attention, based on a previous real lameness that did evoke a sympathetic response. The cardinal fact to remember during this developmental stage of the dog's mind is that there are precise limitations to a dog's intelligence. Monkeys, for example, find detour tests in which they have to get around objects to get to some food very easy to overcome. Dogs find detour tests difficult.

The dog's mind has already been set on a certain course by his breeding, his genetic inheritance, by his mother's and his own hormonal influences and by his relationship with his mother, his littermates and with us. Early experience has set the tone for his future behaviour and he is now at a most active learning stage. Learning goes on constantly. Throughout his life there will be continuing gains and losses in knowledge but now is a time primarily of gains. The principles of learning apply surprisingly well to most mammals. A rat learning to find his way through a maze and Shamu the killer whale learning to toss her trainer up into the air both use the same learning principles. The ability to learn varies, of course, from species to species and has its basis in the evolution of the individual species. Dogs are socially oriented animals and as a consequence, they respond actively to praise and affection. Less sociable animals such as cats respond correspondingly less well to these rewards. The dog can learn in similar ways to the rat and the killer whale and it's worthwhile to describe them briefly here.

1. LEARNING BY OBSERVATION

This might seem obvious but is still one of the earliest and commonest ways that dogs acquire knowledge. One of my dogs will go to the toy box, select an object, bring it back and chew on it. The other, observing what the first dog has done, makes a decision in her mind and does the same thing. This type of learning by observation has been demonstrated in laboratory conditions too. Pups in the juvenile period of development were taught to pull a food cart on a runner by means of a ribbon. Other pups in an adjacent compartment were allowed to watch and learned to do the same thing.

The monkey troop that I mentioned in the Introduction, that learned to wash their sweet potatoes by observing Imo doing so, were far better than dogs at learning by observation, and for a simple reason. Skilled voluntary movement needs a good pyramidal system in the brain. The pyramidal system accounts for almost 30 per cent of the white matter of the human brain, 20 per cent, of it in the primate brain but only 10 per cent of the dog brain.

2. CLASSICAL CONDITIONING

As I've mentioned, I usually distract young pups brought in for their first inoculations by giving them yeast flavoured vitamin tablets. Occasionally there are medical circumstances that result in my seeing a dog many times over a relatively short period of time and each time the dog receives a few vitamin tablets. A not uncommon consequence of my behaviour is that these dogs come in, look at me and dribble. I've unwittingly played Ivan Pavlov and have classically conditioned these dogs to salivate.

Classical conditioning need not be so dramatically obvious. The 'visceral reaction' might not be as obvious as salivation. It might, for example, involve activation of the adrenal gland and the fight or flight response. Hitting a dog with a rolled up newspaper can activate this response. But once you have disciplined a dog several times this way, just the sight of the newspaper can affect him 'viscerally', causing activation of his adrenal-pituitary axis. His mind has been classically conditioned and he experiences a hormonal response to the simple sight of the newspaper. (I know of a Yorkshire terrier, disciplined as a young dog with a newspaper, that as an adult dog developed an insatiable desire to tear newspapers to shreds before his master came downstairs in the morning.) The point to remember about classical conditioning is that the dog's 'visceral' response is involuntary. He has no control over the response. Breeders understand this when they use the same room with the same sights and the same smells each time their stud dog is used for mating. After a few successful matings in that room, just the mere sight of the room can sexually arouse the male. He's been classically conditioned to the breeder's preparations and mating is less dependent on the attractions of the particular female.

3. OPERANT CONDITIONING (OTHERWISE KNOWN AS DOG TRAINING)

Operant conditioning is the behaviourist's term for standard dog training. Other terms are used too, instrumental conditioning, instrumental learning or operant learning. The words 'learning' and 'conditioning' are interchangeable. The most important fact to remember is that dog training is constantly going on whether or not people are actually involved in it. The principle of this type of conditioning of the dog's mind is that a certain action carried out under certain circumstances is followed by a reward. A dog chases a car driving by and the car 'runs away'. Action – reaction – reward. This simple approach to learning is based on work that was originally carried out by B. F. Skinner in the late 1930s in which rats in boxes were 'trained' to do things for food rewards.

Naturally, there are some things that are easier to train a dog to do than others. If I give my dog a tasty yeast tablet each time she barks, I've virtually guaranteed to increase her barking in my presence. Action – reaction. Response – reward. Barking is a behaviour that is easy to learn because it is a natural and frequently used part of normal behaviour. Climbing a step ladder, on the other hand, is more difficult to learn because it isn't part of the normal repertoire of behaviour. This is a behaviour we have to create – to shape – from more naturally occurring behaviours.

Dogs don't need teachers to condition their minds. They don't even have to be in a 'learning' frame of mind. Our role in operant learning, in dog training, is technically speaking, not to be the 'teacher' but to be the 'controller'. Dogs are learning all the time and our objective is to control the stimuli, responses and rewards. We can do so by reinforcing, not reinforcing or punishing the behaviour. Before we go on to further details I should briefly explain what is meant by these terms.

Reinforcement

Dogs are constantly learning in the operant conditioning sense of the word. They learn fastest when their behaviour is consistently rewarded. The rewards themselves will vary. The dog that digs under a garden fence and escapes out onto the street is rewarded by the mental stimulation that comes with exploration. It was fun. He'll do it again. The behaviour is reinforced. Food rewards, praise and petting are simple methods of reinforcing behaviour that we want our dogs to have. Dogs have been trained in the laboratory to bark exactly thirty three times for a food reward. Dogs differ in their enjoyments. To some, exercise is the greatest pleasure. To the more gluttonous, food is the supreme reinforcer. With this in mind, any dog training that we participate in, that we want to control, should be reinforced with a variety of rewards – food, praise, touch or activity. (These will be covered in greater detail later in this chapter.)

Food and physical contact can be called primary reinforcers. Both are potent rewards to the dog. Verbal praise is a less potent reward and can be called a secondary reinforcer. In operant learning, it is always best to use a primary reinforcer first to reward a dog and then to couple that with the secondary reinforcer. An example can be teaching a pup to sit. His meals are potent rewards, primary reinforcers of behaviour. Holding his food bowl directly above his head can naturally bring the pup into the 'sit' position and once he is in that position, he can be given his food. He will soon learn to sit when he sees his food bowl. And if you command 'sit' each time you feed him, this acts as a secondary reinforcer. Soon, simply say the word 'sit' will have him obey without the need of the original primary reinforcer.

The timing, intensity and intervals of reinforcement all have direct consequences on learned behaviour. Reinforcement must be immediate. Giving a dog a reward at the end of training is pointless because you're not reinforcing behaviour immediately. To the dog's mind, a reward must be instantaneous if the behaviour is to be reinforced. The intensity of the reward is also important. If the reward itself is too great, the dog's mind gets in a turmoil and becomes less effective. I see this routinely when dogs are brought to my clinic. Owners try to soothe their pets by stroking them, using 'buzz words' like 'good dog' or even feeding them favourite snacks. In the meantime the dogs are acting scared witless and what the owners are really doing is reinforcing this fear behaviour and with intense rewards.

The schedule of reinforcement of behaviour is also important in modifying the dog's mind during this stage of his development. Continuous reinforcement by which a behaviour is rewarded each time leads to rapid learning. Behaviours that are conditioned by intermittent reinforcement are harder to create, but once created are also more difficult to overcome. Many behaviours become integrated into the dog's mind through initial continuous reinforcement followed up by intermittent reinforcement. A pup, for example, might be allowed to sleep in his owner's bedroom. Well and good. I think this is right and that if you want a sociable animal over which you have constant control, then the pup should not be isolated from his new 'pack'. But the time comes when he is old enough to sleep elsewhere. The initial behaviour pattern was constantly reinforced and readily learned. It's now time for that behaviour pattern to be lost from his mind. This is called extinction of a learned response and occurs most rapidly when a behaviour has been learned through constant reinforcement.

But let's say that you're a softy, that every now and then you still let him sleep in the bedroom. By doing so you are intermittently reinforcing his behaviour and this type of training is far more difficult to overcome or extinguish. The same principle applies to intermittent feeding from the table. It is harder to stop a dog begging food if he is fed titbits intermittently than if he's fed daily from the table. The most long lasting behaviours are created in the dog's mind through variable interval reward schedules. In the war of nerves with your dog, when each is holding out as long as possible, dogs trained to variable interval reward always win.

Types of rewards

In the dog's mind there are degrees of reward and the more valuable the reward, the faster his learning will be. This means that whenever possible and especially during the juvenile learning period we should withold the most potent rewards, food and touch, and only give them at the proper time to suit or elicit proper behaviour. We should also vary rewards as much as possible to reinforce specific behaviours.

Reinforcers are positive or negative. Positive reinforcers are pleasureable. Negative reinforcers are punishing.

1, Food
This is a powerful reinforcer of behaviour. Use it carefully.

2. Touch
Contact is a potent reward for the sociable canine and should be used wisely and carefully simply because different types of touch mean different things to the dog. Gentle petting has a calming effect but lengthy petting, while giving us emotional satisfaction, actually means 'dominance' to the dog and can result in chronic attention seeking from him. Rough and tumble contact will make a dog more excitable and is not condusive to learning.

Voice is a potent form of punishment or reward.

3. Sound

Praise is an excellent secondary reinforcer but must be used initially with a primary reinforcer such as food or touch. The tone of voice, rather than the actual word, is most important. A relaxing tone of voice should be used for positive reinforcement and a harsh tone should be associated with negative reinforcement and punishment.

4. Play

Physical activity is an outstanding reinforcer of behaviour but because it is so exciting, it is often difficult to use.

5. Owner Attention

As leaders of the pack, or at very least, as members of the pack, we often inadvertently reward behaviour simply by paying attention to our dogs. A shout in response to a dog barking can be a potent reward for barking. Acting as doorman to your dog's scratching to go in and out is another suitable reinforcer of this behaviour. And rewarding this behaviour intermittently makes it even harder to extinguish.

6. Chewing

Mouthing activities such as chewing on toys (or chewing up carpets and furniture) can be rewarding by having a subduing effect on the dog's state of arousal, on his adreno-pituitary axis. Urinating and defecating might produce similar awards.

7. Discomfort Avoidance

Avoiding physical or psychological discomfort can be a potent reward. This can be a cruel method of training and physical pain must be avoided except for the most heinous of crimes. Shock collars and high frequency sound collars are high tech methods of negatively reinforcing behaviour. Punishment is not always productive but it does have a place in altering the dog's mind.

Punishment

Physical punishment is an integral part of the development of the dog's mind but it is nowhere near as important as many dog trainers like to believe. Mothers do indeed punish their pups but the consequences, as Erik Wilsson and Per Erik Sundgren in Sweden have observed, are less socially gregarious animals with poorer performance in fetching and other social activities. Physical punishment continues as relationships in the pack evolve but virtually all punishments are related to dominance, not to other behaviours like barking, urinating, digging, exploration or defecation. Natural punishments in the wolf pack can be extreme. Dominance is maintained through ritual displays and wolves rarely fight. When they do however, the fights are savage. The punishment to the loser is traumatic and severe. Through our intervention in breeding, many of the ritualized behaviours to avoid fights have been lost. Although it still amazes me how infrequently dogs fight, they still fight far more frequently than wolves do, and use fighting as punishment in dominance, territorial, sexual, possessive and rivalry disputes.

This type of severe punishment can make the loser abandon his behaviour pattern pretty quickly but it has to be quite severe, a ferocious attack. It also means that physical punishment, at our hands, has to be traumatic to be effective and it is only very rare circumstances that call for this. There are milder punishments that are effective in modifying operant learning.

Physical punishment is sometimes the only way to ultimately get at a dog's mind. (Sheep chasing and killing is such a potent reward that virtually nothing but severe physical punishment can overcome it.) There are harsh drawbacks to its use. If you administer the punishment directly to your dog, he might grow up either fearing you or, at the very least, have conflicting feelings over how he should behave with you.

The object of canine punishment should be to reveal your power, not to inflict pain. That's the way punishment is seen in the dog's mind. Dogs intimidate each other. They can startle each other and induce fear. Mothers can disregard their pups' implorations and simply walk away as they try to feed. These are the better types of punishment to use as negative reinforcers of behaviour.

There are four methods of punishment:

1. Physical pain
Choke collars, spike collars and shock collars all work on this principle. Striking the dog does as well. As I've mentioned, physical pain should be reserved for the most overwhelming of problems such as sheep chasing.

2. Intimidation
Grabbing the delinquent by his scruff, giving him a shake and an eye to eye stare while admonishing 'Bad dog!' is effective intimidation. Commanding him to sit and just staring at him can be good intimidation too.

3. Fear
A startle can frighten the dog and if it's used in association with

misbehaviour, it can act as a punishment. This is often called aversion therapy. You create something unpleasant that the dog associates with his behaviour and he ceases to carry out that activity. Shaking a tin can with coins or bolts in it, whistles, high frequency sound or simply a harsh voice are all forms of fear punishment. Remember, dogs are always learning even when we're not teaching. All of these punishments can modify behaviour even when we're not around. A shutter suddenly slamming shut as my older dog walked down the street made her unwilling to walk down that street again.

4. Time out
This is a favoured form of punishment, based on the social gregariousness of the dog and his need to be with his pack. The dog is instantly disregarded when he misbehaves. At the instant of wrongdoing you go from the room, leaving him alone. This punishment should last for at the most only a few minutes. The object is not retribution but surprise. It's a theatrical gesture based on the mother walking away from the pup that wants to suckle. Ostracism is a severe punishment and very effective if used properly.

All forms of punishment must be associated in the dog's mind with his behaviour. It should be administered at the time of wrongdoing or within a second of it. Punishment after the event is pointless and even counterproductive. You might get the 'guilty look' but that's fear of you, not guilt over the misdemeanour. Punishment works best if it's given as the misbehaviour begins. A shout as the dog is about to jump up on the table is effective. Shouting at him after is not. Punishment works best if every misbehaviour is consistently punished. The dog that jumps up on the dining-room table in your absence won't respond well to punishment.

Some forms of punishment might be disciplinary to you but actually rewarding to the dog. Hitting your dog might seem to be a form of punishment but some animals crave any type of physical contact and it can in fact be a reward!

The dog's response to punishment varies according to the make-up of his mind (whether he is tolerant, nervous, excitable), his previous experience with discipline, the strength and timing of the punishment and the relative dominance of the dog and owner. Punishment has severe drawbacks. It can produce visceral responses. Some dogs will urinate, defecate or empty their anal glands under even benign discipline. It can interfere with learning or interfere with already learned behaviours. It can produce frustration and anxiety in the dog's mind and produce excessive fearfulness and the associated aggression that goes with it. Punishment given inconsistently can create a condition of the mind called learned helplessness, a neurotic state in which the dog actually learns to be helpless. This is easier than you might think to induce in a dog. Simply calling your dog's name to praise him and calling your dog's name to punish him can lead to this neurotic state.

The dog learns through observation, classical (Pavlovian) conditioning and operant (Skinnerian) conditioning. It gets reinforced by reward. Learning takes place whether we are involved or not and there are constant gains and losses from the dog's mind. These gains and losses can be affected in several ways.

Shaping
We can shape a dog's learning by initially rewarding a naturally occurring behaviour such as following an airborne scent and then 'shaping' his behaviour by only rewarding him for following a certain scent. This is how scent dogs are trained. Rewards are first given for using their noses and ultimately given for using their noses to scent out drugs, people, mould in lumber yards or whatever is required.

Extinction
As has already been mentioned, if a learned response is not reinforced then it's gradually lost and the frequency of the behaviour falls to its naturally occurring level.

Chaining
Dogs can learn to carry out a sequence of events in order. The last behaviour is learned first and reinforced with a primary reinforcer like food or touch and then with a secondary reinforcer like voice. Then the preceding behaviour is learned, giving the food reward for this and the verbal reward for the final behaviour. The chain continues this way.

Habituation
Dogs have the ability to get used to neutral (non harmful and non rewarding) stimuli. This is called habituation and occurs in two ways, through constant exposure to the stimulus or through a gradual introduction to it.

Flooding – habituation through constant exposure
If a dog's mind is continually flooded with something mildly fearful like traffic noise, he learns to disregard what in other circumstances would be frightening and to relax in the presence of that stimulus. City dogs constantly learn through flooding and often develop 'street sense' as a result.

Systematic desensitization – habituation through gradual exposure
If a dog sees a cat at a distance one day, then at a closer distance
another day, and gradually over a time at closer and closer distances,
his natural aggressive or fearful behaviours might be altered. Getting a
dog used to something that causes fear by gradually exposing the dog to
the fear inducing stimulus (like noise) is a common method of modifying
a dog's behaviour.

Although the difference between classical conditioning and operant
learning might seem a little dry and academic, it is still important if we
are to understand the continuing development of the dog's mind. Behaviours that develop through this juvenile period and on into the adult
stage of his life are conditioned in these two ways. As we now become
an integral part of the dog's life and are so influential on his mind, it's
important that we know the difference, between these two learning
methods so that we can modify his behaviour in the best possible way.
By understanding that classical conditioning does not involve a reward
principle but that operant conditioning does, we have a better likelihood
of success in altering the dog's mind. Take house training for example,
the behaviour that is initially of most interest to any of us with pets.
Various internal bodily sensations such as a full bladder, combined with
external stimuli such as the smell of urine are the influences that cause
the pup to urinate. House training involves classically conditioning the
dog's mind to associate these stimuli with the outdoors rather than the
indoors. No reward is necessary. And similarly, punishment, although
satisfying to the owner, is pointless for the dog because this is not an
operant conditioning situation. It's a problem involving classical Pavlovian conditioning.

To effectively mould the dog's mind, our task in training is to arrange
as often as possible that the pup associates the various stimuli of the
outdoors with his unconditioned stimuli such as a full bladder. In
practice that means anticipating when to take the pup outside. If he
starts sniffing, he can be under the influence of unconditioned stimulation (the smell of urine). When he wakes up, he can be under the
influence of unconditioned stimulation (a full bladder). Similarly, after
he eats or plays, these stimuli become dominant. And just as Pavlov
classically conditioned his dogs to salivate at the ring of a bell, we
classically condition our dogs to urinate and defecate to the stimuli of
the outdoors. My older retriever was gutter trained when I lived in the
very heart of London where cars were parked bumper to bumper. There
was so little space for her that her head was frequently under the
bumper of a car as she urinated. Today we live on a less crowded street

but my dog has been classically conditioned to urinate in the gutter only in close proximity to a car and will only do so if she is virtually touching a vehicle. House training is not a battle of wills. Reward can play a part in it. Praise and affection when the pup eliminates in the proper place is effective too because it cements the bond between the dog and you. Elimination, however, is a classically conditioned behaviour. Extinction or punishment will not work in modifying it. If we understand the difference between classical and operant conditioning, we have a greater likelihood of influencing the dog's developing mind.

Classical and operant learning develop to maturity from the end of the dog's socialization period, through puberty, to his eventual emotional maturity at around two years of age. The anatomy of this learning is still poorly known but involves all aspects of the mind. The wiring of the brain, hormones and the senses all play a role. Learning appears to take place in two stages, the formation of short term memory and the consolidation and formation of long term memory. Shock or even general anaesthetics can alter short term memory and it's postulated that short term memory is an electrical event. Chemicals are undoubtably also involved. The protein scotophobin, injected into naive mice, causes them to learn specific tasks more rapidly. It is not too far-fetched to say that, theoretically, one day it will be possible to inject a 'sit-stay' protein into dogs, but in the meantime we can rely on our understanding of classical and operant learning. The 'learning' centre of the brain is throughout the cortex but, to complicate matters, classical learning doesn't require the cortex. Additionally, some parts of the brain are more important for some particular types of learning than are others. The hippocampus is more important for learning passive avoidance rather than active avoidance. Dogs can learn to control their autonomic nervous system through the hippocampus (to avoid receiving electric shocks), a method of learning to control biofeedback.

It also seems that the actual number of neurotransmitter containing vesicles at nerve synapses might increase with learning. That's an exciting find because it means that the dog's mind is receptive to learning as long as it isn't degenerating, that even after all wiring is in place, there is still capacity for brain development that is actually stimulated by learning.

Over the coming months the social development of the dog's mind continues. His methods of communication become more precise and refined and the expressions of his emotions mature. He will express his emotions through body language, vocalizing and marking behaviour as discussed in Chapter Five. The dog's social life and organization depend upon these signalling systems, both in his relationship with other dogs and with us. He reflects his mood in his body carriage, the position of his ears and tail, the articulation of his limbs. He signals his feelings with his voice – conversation, satisfaction, settling down, fear, aggression, alarm, dissatisfaction, pain. He marks with his body waste products or simply by scratching the earth (or floor or walls).

Dogs remain the gregariously social creatures that they are throughout their lives but now often in the absence of other dogs. We make quite good dog substitutes but fail in some ways. A dog will, for example, eat more in the presence of another dog. One facilitates the behaviour of the other. Some pet owners almost instinctively understand this and will pretend to eat a dog's food in order to coax it to feed. Social facilitation might be due to competition or increased emotional activity but, in either case, life becomes more exciting. When running speed has been measured, dogs running in pairs along a 190 foot runway always ran faster than dogs running singly.

The day to day companionship between dogs has a direct effect on their attitude towards us. A study of the dog's emotional response to humans showed that in paired dogs, their responses were similar far more frequently than chance would allow. They synchronized their behaviour.

As the dog's mind continues to develop, a problem occurs. Even though we have selectively bred down many of the 'wild' traits in dogs, they haven't been eliminated. Hunting, tracking, stalking, biting, shaking, killing. These are all normal canine behaviours but where is the outlet for these natural tendencies and basic instincts in companion dogs? David Mech, the wolf behaviour naturalist, has described how 'behavioural synchronization' is a pack trait that has evolved for the collective good. Unity and coordination, says Mech, are key concepts in

Dogs need mental stimulation.

pack behaviour. Pack members should all rest at the same time and be active together. If the pack members are all full of energy, then the group is a powerful hunting unit. This is the unwritten law of all canids.

Dogs are faithful to this law which is why they rest when we rest, want to eat when we eat and want to be active when we're active. My dogs are as good examples as any. If someone is in bed, they're there snoozing. If food is available, they hover just out of sight. If I have rough and tumbles with the kids, they want to join in. If I'm going for a walk in the woods, they're both at my feet with leads in their mouths, hip flasks slung over their backs and shotguns under their arms. They coordinate with me, the leader of the pack.

Anders Hallgren in Sweden has written clearly of our need to understand this aspect of the dog's mind. Because of the unique nature of the relationship of the dog to us, he is dependent on our initiative for his mental and physical activity. When we fail in this area, we are confronted with the behavioural problems that all too frequently develop. The dog's mind has evolved to be active in cooperation with other dogs. Hallgren says that we should bear this in mind when it comes to understanding their behaviour and preventing behaviour problems. He says that we should stimulate the dog's mind in similar ways to those in which it is naturally stimulated, but channel this energy into useful activities which he divides into, nose work, problem solving, learning and balance.

We have to think like dogs, get into their minds, if we're going to be successful. It's dead simple to make mistakes and drive a dog crazy. Teaching him as a pup that calling his name is associated with petting and friendliness and at the same time calling his name to discipline him when he messes in the house is, as I've mentioned, a sure way to

If a dog's natural energy is not properly channelled, he can indulge in destructive behaviour.

produce a mixed up mutt. Our involvement in the development of the dog's mind should be fun and before we look at specific social behaviours and problems it might be best to list some rules of canine learning.

Rules of Canine Learning

1. Normal dogs can learn at any age up to the time their mental capacities start to deteriorate in old age.
2. They learn best through patience and with suitable rewards. Rewards must be given with or within less than a second of the desired response from the dog.
3. Intermittent rewarding of desired activity produces behaviour that is more resistent to extinction.
4. The value of the reward should be appropriate for the desired behaviour. Dogs value rewards in different ways. Find out what is most valuable to your dog and use the rewards appropriately.
5. Learning should be enjoyable. Spend ten minutes two to three times each day. Sessions should be separated by several hours. Tired dogs do not learn easily. Mental activity is more tiring to the dog than physical activity.
6. Learning should take place in a quiet environment. Once the correct responses have been made in that environment, you can move to more stimulating environments and repeat the processes.
7. Every dog should be trained to come, sit, stay, down and down stay in that order (see appendix). If a dog fails at any level, do not punish but simply go back to the previous level. Always finish training sessions on a positive note.
8. Only use your dog's name to get his attention. Use one word commands in training.
9. Once your dog has learned commands from one person have him learn the same commands from other members of the household. In this way he learns that in a dominance hierarchy, he is beneath all the humans in his pack.
10. Punishment can be counterproductive. Use it with care. Natural punishments using your voice and stare can, however, be effective.

All developments of the dog's mind derive from combinations of instinct and learning. They are dependent upon genetics, hormones, senses and all the external stimuli that the dog receives from his mother, his littermates, from us, from other dogs and animals and from his environment. This is nowhere clearer than in the context of aggression, such a large subject that it deserves its own chapter.

Chapter Eight

Social Behaviour – Aggression

There are hundreds of research reports on the subject of aggressive behaviour in dogs but there is still no acceptable definition of aggression. The word often conjures in our mind maliciousness, nastiness or vindictiveness but that is not what aggression means in the dog's mind. Aggression takes many different forms. Dominance aggression is primarily influenced by genetics. Inter-male or sex related aggression is hormone related while other forms can be influenced by the environment (moving objects) or are learned in the operant learning mode. Some types of aggression are simply related to the excitement level of the dog's mind.

Aggression.

Aggression can be subtle. It doesn't simply involve going for the jugular. Ritual forms of aggression are finely developed in dogs and we shouldn't look upon this behaviour as sinister, pathological or wrong. It can certainly be unpleasant to our minds but can often involve simply a look or a posture rather than a bite. Bites, however, do indeed occur and other dogs aren't the only recipients of these bites. Several million serious dog bites are reported each year in the United States. That's serious dog bites. The number of less serious unreported dog bites is unknown.

Aggression involves threats through posturing, barking and growling or just staring and can be offensive or defensive. Dogs can chase and bite but equally they can flee and bite. Aggression is influenced by sex, age, size, hormonal status, territory, personal distance, dominance/subordinance hierarchy and the outcomes of previous encounters with individuals. To the dog's mind, many aggressive encounters are a form of competition over either a resource (e.g. sibling rivalry) or are an opportunity to enhance his genetic potential for survival – to produce offspring. The resource at the centre of competition is sometimes obvious – food, shelter, territory. But just as often, it's difficult to see exactly why the dog is being aggressive. In these situations the pay-off comes later, such as climbing up a rung in the dominance hierarchy. This is the most common type of aggression and a good place to begin looking at why dogs behave in this way.

1. DOMINANCE AGGRESSION

Dogs are pack animals. They simply don't expect equality. Their minds are not wired up in that configuration. The dog's natural genetic predisposition is to find his place in the pecking order, the dominance hierarchy. For some dogs this means adding notches to his gun, trying to be top dog. Dominance aggression, according to Victoria Voith, is a problem that is usually shown by male dogs between two and two and a half years of age. The curious fact is that most male dogs reach puberty much earlier. Their testosterone surge occurs at between six months and a year of age yet the apparently sudden, unprovoked aggressive attacks of dominance aggression, usually against members of the immediate family, frequently don't occur for another one to two years. The onset of these attacks coincides with the time of puberty in wolves. Selective breeding has made dogs precociously sexually mature at an early age but emotional maturity takes longer.

Dominance aggression can be provoked in a myriad of ways and is a pack problem. As we humans make up most dog packs as far as the dogs are concerned, it is usually directed at us. This type of aggression can be provoked by simply disturbing your dog, such as awakening him or moving him or ordering him to move from his resting place. It can be provoked by approaching his food, his 'favourite person', or his resting area even if he's not in it. Dominance aggression can be stimulated by petting, by having collars and leads put on or off, by being stared at or disciplined, by grooming, nail cutting, towelling or even by meeting in a narrow passageway where the dog feels he has right of passage first. Discipline, either verbal or physical, can provoke dominance aggression. Just standing over a dog can be interpreted as a threat to his position and can invoke a dominance response.

To complicate matters a little more, a dog might show dominance aggression only in very specific circumstances. It might occur only in one specific place or at a certain time. This is common with grooming. A dog might 'fight' grooming if you normally carry it out in the kitchen at lunchtime but if you unexpectedly do it first thing in the morning outdoors, he might not show his usual dominance aggression. Some dogs can be psychologically dominant but physically submissive. This is particularly true of some toy terriers, dogs that enjoy being carried and tickled but only in their own time, biting their owners if the humans get out of line.

Approximately one out of every five aggression problems brought to the veterinarian for attention is one of dominance aggression. What happens is that the dog perceives his owner to be weak. If my dog comes over and asks to be stroked, and I stroke her, I'm rewarding her initiative. But at the same time I'm telling her that I obey her commands. Dominance can be as subtle as that. When dogs interpret our behaviour as a reaction to theirs, their natural inclination is to feel more assertive and dominant. And if their genetic predisposition is towards active dominance, some vie for the 'alpha' position in the household pack, and often achieve it. In nature, rivals for the 'alpha' position in the wolf pack will eventually fight for dominance and the physically strongest takes over. The loser assumes a lower status. When we try to win our dog's affection by responding to their demands, we can unwittingly be creating a feeling of dominance in their minds.

Dominance aggression is both inherited and learned. Although 85 per cent of clinical cases occur in males, it is not puberty related. The hormonal influence on this behaviour probably occurred near birth when the male pup's brain was 'masculinized' by a surge of testosterone. This is why castration has little beneficial effect on correcting this unpleasant behaviour.

Direct eye contact, erect ears and baring of teeth are all postures of dominance aggression.

Dogs exhibit their dominance aggression in many ways. There are the typical dominant postures such as 'standing over' an object – guarding it, direct eye contact – the stare, tense, rigid posture with erect ears and tail, growling, teeth baring, snapping, biting. This type of aggression is usually directed at members of the family or other members of the dog's social group and is more likely to occur in households with kindly non-authoritarian type owners who allow their dogs to get their own way or treat them as equals. An interesting aspect of the problem is that dominant aggressive dogs rarely attack very submissive people or small children. In their minds they only perceive more forceful people as threats to their social position. (Children do, however, get bitten more frequently than adults but this is usually fear related or competitive aggression.)

Although his genetic inheritance, the early masculinization of his brain and the size of his weapons all play a part in dominance aggression, learning is also a strong component. This is fortunate because it means that treatment is possible. Allowing a dog to get his own way permits and potentiates dominant behaviour. Backing away from a growl, allowing the dog to win tug of war games, letting him wander freely, allowing him to jump up on people – permissiveness creates problems but this aspect of dominance aggression can be overcome through retraining.

Prevention

Dominance aggression can best be avoided through careful selection of a dog in the first place. This type of aggression is reported in certain pure bred dogs more frequently than in crossbred dogs, specifically in English Springer Spaniels, English Cocker Spaniels, Rottweilers, Dobermanns and rather surprisingly Lhasa Apsos. Hounds, on the other hand, rarely show this behaviour. We know that puppy testing is not a predictor of future general behaviour, but pups that test positive for dominance do grow up to be dominant. Avoid these pups and avoid pups from parents that are known to be aggressive in a dominant way.

After selecting your pup, dominate it while it is young and impressionable. For example, the dog should eat on your terms not his. Train him by occasionally taking his food or toys away and then returning them. Brush and groom him and trim his nails frequently. Never let threats go unchecked. If he nips and mounts, he needs to be chastised. Punishment as we know is a poor correction for learned behaviour. Extinction works better. Simply avoid the situations that lead to his learning dominance. Avoid tug of war and other rough and tumble or chasing games. Games increase excitability, physical competitiveness and aggressive dominance. They also encourage other unwanted behaviours such as chewing and biting.

Condition your dog early to the lead and start obedience exercises as soon as he comes into your home. Teach him to sit on command before he is fed and before he is stroked. The absolute law is that he must do something for you before you do something for him. In that way, he learns and hopefully remembers forever that you are the leader of the pack and only you have the right to be dominant. By 'you' I mean all competent members of your family, his 'pack'. Everyone in the home should participate in training. First of all, it's fun. It's fun for your dog too. He will enjoy knowing where his place is in the pecking order.

Treating dominance aggression

Your primary goal in treating dominance aggression is to avoid having people injured while at the same time getting your dog to assume submissive and non aggressive behaviours in the circumstances where he previously thought he was top dog, where he actually WAS top dog. Whenever possible avoid the specific situations or stimuli that you know provoke dominance aggression. For example, if brushing your Yorkshire terrier provokes him to attack you, temporarily stop grooming him.

The dominant dog must learn that from now on nothing is for free.

The first object of treatment is to disorient his mind, to force him to reevaluate his relationship with you and your family. Be consistent and have the most dominant member of the household (the pack) take over training. This, incidently, is not necessarily the man of the household. In practice I frequently see dominance aggression problems in pets caused by putty soft husbands coming home from work and undoing all the good discipline instilled during the day by their wives.

Retrain, or train your dog for the first time to sit, stay, come, down, down stay and heel, in that order (see appendix). The down stay is very important because this is a subservient position. You may need professional help in teaching these routines but do not send your dog away for training. The aggression problem is between your dog and his pack. He has to learn to be subservient to you, in your home, not to a stranger in a distant kennel.

As preconditions to training, remove all your dog's valued assets, his toys, bones, even his blanket. This is serious business. Do not give in to your human emotions. Think only on his terms. You want to be in command and will do anything to usurp the power he has usurped from you. And remember, aggression is the most common cause of death in young dogs. They don't die as a direct consequence of aggression. They are destroyed by us because of it. Avoid confrontations in which your dog may win. Any winning on his part puts retraining back considerably. And don't physically punish a dominant aggressive dog. Hitting him can simply stimulate further dominance.

Small dogs are easier to treat. We can show dominance by scruffing them, picking them off the ground or shaking them. Large dogs can be grabbed on each side of the neck and given a lift, shake and stare, but be careful. This can be quite dangerous with a very dominant dog. Only do this if you are certain that you are firmly in control of the situation.

Although castration has little effect on dominance aggression on its own, this problem often occurs in conjunction with intermale aggression. The consequence is that there are some instances where castration is beneficial. Similarly, treating with the female hormone progesterone can be useful but only in conjunction with retraining. Hormone treatment on its own will not solve this problem. The good news is that through behaviour modification and hormonal alteration, the mind set of the dominant aggressive dog can be altered. The bad news is that the problem can never be eliminated. That's because the mind set is genetic in many animals and therefore unapproachable for modification. The prognosis therefore is guarded. There is a risk of injury in any household with such a dog.

2. POSSESSIVE AGGRESSION (INCLUDING COMPETITIVE AGGRESSION AND SIBLING RIVALRY)

Many years ago Dr Theodore Zahn noted that when pups were taken out of a pen then put back in, 'jealousy' was provoked from others in the litter. He thought this occurred only when definite dominance/ submission relationships had not been established, but it can be perpetuated through the juvenile period and into the adult stage of the dog's life. Sibling rivalry occurs when two dogs are so similar that they find it difficult to determine which is dominant. It is a form of dominance aggression and almost routinely the problem is exacerbated through our intervention. This learning problem occurs most frequently in dogs that resemble each other in size, sex and age and is usually triggered by competition for food, toys, a sleeping place or our attention. Two delightful male Golden retrievers aged ten and six years started fighting with each other each time the owners came home, a simple case of sibling rivalry. The older dog had always been the dominant one but now the younger felt he should be top dog. He got upset because the owners always stroked the older dog first. The owners were the unwitting cause of the problem. Once they acknowledged the change in status between their dogs, and started stroking the younger dog first when they returned home, the fights, and these were vicious fights, stopped.

Dogs don't expect to live in equality with other dogs or with us for that matter. Democracy is a misplaced ideology in the canine world. Their minds work differently but we find this a difficult concept to accept because it's alien to the way most of us think. Kind owners who punish bully dogs and elevate submissive ones create conflict in their dogs minds because, to their thinking, what we are doing is unnatural. Jealousy over attention from the owner, possessiveness over a toy or a bone, rivalry over who sleeps in the favourite spot, these are all manifestations of problems in dominance behaviour where a hierarchy has not been properly established.

Prevention

Possessive or competitive aggression is a form of dominance aggression and as such is primarily under genetic influence, but one aspect of it is more open to prevention and that is the learned component. Competitive aggression is less common in breeds that have evolved as pack hounds, beagles for example. It is more common in terriers that have been bred to hunt singly. If you keep fox terriers, for example, you are almost assured of there being competitive aggression and sibling rivalry between them. Possessive aggression is best prevented by choosing your dogs carefully and showing dominance over them when they are young. Possessive aggression is the variation of simple dominance that dogs most frequently direct against children. In the dog's mind, it's a true case of sibling rivalry. He might know that adults are leaders and that he leaves their food alone but at the same time sense the lack of authority of young children. In this dog's mind, children become competitors for food, or affection or play from the leaders of the pack, the parents, and he shows his rivalry by being aggressive with the child. The best prevention then, is to train the dog from the moment you get him that all members of the household are dominant over him. That means that children who are old enough to do so should be involved in all training. Even five or six-year-olds are old enough, under supervision, to learn how to command a family dog to sit and stay. Remember, however, that dog training isn't child's play. Children aren't really old enough to be completely responsible for the family dog until they are physically and emotionally well along the way to maturity, as young as ten years old for some children, and never for others.

Our responsibility during the juvenile period of the dog's mind is to determine which of two or more dogs is likely to be dominant over the other(s) and to then give subtle rewards to that dog, rewards like the first pat when you return home. At the same time, he must remember that you are dominant over him. The submissive dog, in the meantime, gets ignored. This attitude on our parts, hard as it may seem, lengthens the distance between the two dogs and dramatically reduces the likelihood of future fighting.

Treating possessive aggression

Treatment is the same as prevention. If the form of possessive aggression is sibling rivalry, you, the owner, must decide which dog is naturally dominant and which one is submissive and then reinforce the natural situation.

Acknowledging a change in dominance hierarchy between dogs is a simple solution for overcoming sibling rivalry between dogs but is a disastrous recipe if the competitive aggression is directed at members of the family, especially children. Under these circumstances follow all the rules for treating simple dominance aggression.

3. FEAR AGGRESSION (INCLUDING PAIN INDUCED AGGRESSION)

The very opposite, the antithesis of dominance aggression, is fear induced aggression. Dominance aggression is primarily genetic and difficult to alter. It is offensive in the behavioural sense of the word. Fear aggression is defensive and is primarily a learned behaviour. Because it is learned, it is more treatable.

We already know from the early work of canine behaviourists that if a pup is not properly socialized to his environment, if he does not experience sights, smells and noises during his formative socialization period, then he can develop a fear of the unusual. Fear is the natural self-preservation response of all animals to new or unusual situations and is the most common response of all captive animals towards humans. Perhaps surprisingly it is also the most common type of aggression that pet dogs exhibit. One out of every four cases of aggression that is brought to the veterinarian's attention is fear aggression or its equivalent pain induced aggression.

The dog's response to fear or pain is a mixture of physical, physiological and emotional, and although many types of fear induced aggression

Fear aggression is the most common cause of dog bites of children.

have their origins during the socialization period from seven to twelve weeks of age, a dog can develop fear behaviour at any age.

The body posture of fear can be a mixture of subservient gestures and aggressive ones. The ears are plastered back on the head. The tail is held very low or between the legs and will usually wag in short quick movements. Towards humans that the dog knows there is a submissive grin, but at the same time a degree of open-mouthed threat with a retraction of the lips and a show of teeth. The dog's back might be arched and his head held low. He might even show licking movements with his tongue, a submissive gesture that, together with the retracted lips, shows the ambivalence of his feelings in certain situations.

Fear aggression is the most common cause of dog bites of children. Boys over five years old are bit twice as often as little girls and five to fourteen-year-olds are bit the most. (In seasonal climates, bites occur most frequently in the spring and summer.) These dogs bite children for many reasons. If they have not been properly socialized to children, they can think of them as a new and unfamiliar species. To the dog's mind our children are quite different to us. Prepubertal children smell different as well as being smaller. They move in a much jerkier fashion.

Fear aggression can be caused by a previous painful experience with a child when, for example, the dog's hair was pulled. Some dogs can even develop a fear of children because they have previously been disciplined or scolded in the presence of the child and associate the child with that experience.

Fear aggression occurs almost equally in males and females and is not hormone related, although there is a slightly greater incidence in intact males and spayed females than in castrated males and intact females. The behaviour is reinforced by learning. The dog shows fear and snaps at someone who approaches him. That someone backs off and the dog learns that snapping when fearful works.

Although many types of fear induced aggression have their origins in early learning, there is also a strong genetic component in certain breeds and individuals. Remember Allegheny Sue and her neurotic pups? That was an instance of inherited and maternally imprinted fear behaviour. Allegheny Sue was genetically pre-wired to be submissive and fearful and passed this genetic potential on to her pups. This type of fear usually develops in the pup at some time between three and ten months of age. It is completely unrelated to inherited dominance aggression.

Inherited fear aggression is a more serious problem if it occurs in one of the large defense breeds and it does in the German shepherd. Konrad

Lorenz calls these dogs 'angstbeisers', an apt description of their behaviour. All dogs inherit a certain reactivity to sounds, to movement, to touch and it is an exaggerated reactivity to these types of sensory stimulation that provokes fear aggression in certain breeds such as the German shepherd.

Fear aggression frequently occurs on its own but can also accompany other signs of fear, barking, trembling, pacing or hysteria. In situations of extreme fear, some dogs will empty their anal glands, urinate and defecate all while trying to bite. I routinely see a fearful Chihuahua that behaves in this way. All he has to do is look at me and every orifice on his body opens. His pupils dilate, his heart rate increases, his blood pressure goes up, he attempts to bite and he urinates and defecates.

Prevention

Many forms of fear induced aggression can be avoided through proper socialization of pups. A young pup will adapt to his environment and will grow up not being fearful of the noises, smells and sights he encounters as long as they are assimilated into his brain at an early age. A pup should be exposed to children, postmen, cats, other dogs, traffic noise, crowds, elevators – everything he might conceivably encounter later on in life. If the dog's mind is barraged with sensory stimulation at an early age, he is less likely to be fearful later in life.

Inherited fear aggression is less easy to prevent, especially the global fear of all stimuli that some dogs exhibit, but this too can be ameliorated through proper socialization. Careful selection of a pup is important. The cowering pup in the corner might be a heart tugger but also has the potential to become a fear biter because of inate submission and fear. The principles of learning apply to the prevention of fear aggression in all dogs whether they have the genetic predisposition to bite or not. Some punishments of submissive dogs can turn them into fear biters.

Treating fear aggression

Because so many causes of fear aggression are learned, it is possible to alter the dog's behaviour through desensitization and counter-conditioning, using various types of rewards, food, touch, play and grooming. There are several principles of treatment.

1. Teach sit-stay for a food reward several times daily until your dog responds with a sit for several minutes.
2. Identify the exact cause of fear aggression. For example, it might

not be all children. It might just be one child, or even one child wearing certain clothing.

3. Identify the exact circumstances under which the aggression occurs. This is important because in re-training, we want to recreate the circumstances but in a less fearful way.

4. Identify the circumstances in which your dog does NOT show fearful aggression. He might be fearfully aggressive to one person but not to another.

5. Never use punishment to treat fear aggression. It will only make the situation worse.

6. Flooding is not a sensible way to treat fear aggression. Although constant exposure to the fearful stimulus, children for example, might eventually habituate the fearful dog to their presence, it is too dangerous and unreliable a technique to use.

7. Desensitizing by exposing the dog to the fearful stimulation at a lower intensity and giving powerful rewards (food and touch) for NOT showing fear aggression is the best method of training. Remember, stroking your dog to calm him when he shows signs of nervous fearful aggression exaggerates the problem. You are actually reinforcing his aggressive behaviour by rewarding his fear with contact comfort.

8. Training sessions should be fun. Each session should last between fifteen and twenty minutes and be performed twice daily.

9. Command your dog to sit-stay beside you and have the person (or other dog) that does not provoke fear aggression walk towards you. If there is no sign of fearful behaviour at five metres, the dog is rewarded with food. If there is still no sign of fear aggression when the person raises his arm, the dog gets another food reward.

10. Repeat this procedure so that the person gradually moves closer and raises his arm, finally touching the dog. If and when a distance is reached where the dog shows signs of nervousness, do not give him a reward, break off activity for ten minutes during which he should be ignored, then resume the training once more at the previously successful distance. Remember: always finish a training session on a positive note.

11. Perform these activities in the least threatening environment with the least threatening person. Once the dog is systematically desensitized in this way, repeat the entire procedure but make it more threatening to the dog by having your friend wear strange clothes or walk with a limp or with a cane or in any other unusual fashion. Once more, remember theatricality might make you feel foolish but it works marvellously on the dog's mind.

12. Once the dog is retrained under these circumstances, return to the same environment, but this time with someone who is a stranger to your dog. If you are desensitizing your dog to another dog, use a placid stooge, preferably an easy going female who will not be distressed by any fear aggressive behaviour from your dog. Repeat the entire procedure until there is no fear response to the strange person or dog at close proximity.

13. Finally, take your dog to what he considers a more fearful environment and repeat the above series with a familiar person or dog and then a strange person or dog.

Time consuming, isn't it. But it works. The most important rule to remember is that prevention is far easier than cure and of the many types of aggression, this one is most preventable. If the dog's mind is exposed to the stimuli of life early, when he is in the socialization phase, you can save yourself a lot of time and worry in the future.

4. PROTECTIVE AGGRESSION (INCLUDING TERRITORIAL AND MATERNAL AGGRESSION)

Dogs protect what they consider to be theirs in many ways. Biting the postman, chasing cars, guarding toys, these are all manifestations of protective aggression and are almost as likely to occur in females as in males. I've used the word 'almost' intentionally because although I am describing different types of aggression – dominance, fear, protective – clinically or behaviourally speaking, aggressive behaviour is not so neatly segregated. Protective aggression can have a dominance component to it and this is more likely to occur in male dogs. On the other hand, protective aggression can also be motivated by fear and, at the opposite extreme, it can manifest itself as maternal aggression, the female's protection of her litter or even of her toys or possessions when she is experiencing a false pregnancy.

Most forms of aggression are multi-dimensional and this is most true with protective aggression. As an entity, it is a common problem, as common as competitive aggression, accounting for one out of every five cases of aggression brought to the veterinarian's attention for treatment. In fact, the four types of aggression I have mentioned, dominance, fear, competitive/possessive and protective, account for four out of every five cases of aggression in dogs. (The four other categories, inter male, predatory, learned and idiopathic aggression, account for the remaining 20 per cent of cases.)

In other species, protective aggression is usually called territorial aggression. Pet dogs do, of course, sometimes aggressively protect their home territories and that can mean the house, the garden, or the car. But they can also, just as aggressively, protect the owner, the children or other animals whether or not they are on their own 'territory' and that's why I prefer the term protective aggression. A unique form of this type of aggression occurs in the female, the only type of aggression that is specific to that sex, and that is maternal aggression.

Maternal aggression is influenced by the female hormone progesterone. Curiously there is no need for pups in order for a bitch to behave in this way. Naturally, if there are pups, her behaviour is understandable. Maternal aggression protects the young, preserves her genes. It can frequently occur without a display of threat. The female simply launches a preemptive attack. Selective breeding has dramatically reduced this behaviour in dogs but it can still occur. Nursing mothers might take their litter to a safe den, under a bed for example, and refuse to let anyone handle the pups.

More frequently, however, the maternal form of protective aggression occurs when there are no pups! Females become apparently possessive of objects and fiercely defend them. And although on the surface it appears that the problem is one of competitive aggression for a favoured object, in fact, hormonally speaking, this is a pure form of maternal aggression, protection of the surrogate pup.

As we already know, dogs are unique among domesticated species in that after ovulation, there is always a surge of the hormone progesterone in anticipation of a pregnancy. Progesterone production lasts for two months, the length of pregnancy and is the hormonal precipitating factor for maternal behaviour, including maternal aggression. A slipper or a toy is, to the dog's mind, a surrogate pup and can be stoutly defended. I recently had a client come to the clinic wearing different shoes. When I commented on the fact, she simply told me that her German shepherd had recently finished her season and was 'nursing' her shoes under the bed. When this happens she 'jollies' her dog with sweet talk. 'Nice shoe,' she cooes. 'Pretty shoe,' she intones. And, given time, she can touch her shoes, stroke them and finally pick them up and wear them. The morning I saw her she simply hadn't the time to jolly her shepherd into giving up her 'pups'.

Protective aggression is more serious when it is constant. Maternal aggression lasts only two months during the false pregnancy stage in bitches or up to two months longer if a litter is actually produced. Protective aggression in which a dog protects his home, his car, his garden and his family is, of course, one of the strongest values of canine companionship, but it can often become troublesome and must be controlled. This type of aggression usually first shows itself around puberty in males and females. Territorially aggressive dogs are often friendly on neutral territory such as at the veterinary clinic but are ferociously vicious on their own territory. Both dominance aggression and fear-related aggression can be components of the behaviour.

Generally speaking the intensity of the territorial or protective drive is inherited. It's genetic but this basis is augmented through learning. This is why the German shepherd, for example, makes such a magnificent guard dog. It has the genetic predisposition to protect and is a good learner. Operant conditioning enhances the behaviour and, as with many forms of dog training, we don't have to be involved for the dog to learn. Postmen are classic examples. The postman walks up to the door, the dog barks aggressively and the postman walks away. Action – reaction. Response – reward. To the dog's mind, it was his aggressive display that caused the postman to retreat.

From our point of view, protective aggression can be a good thing. I like the fact that if my dogs hear noise outside my home, they bark. It is a natural behaviour that I have been happy to see develop. I wouldn't be happy if they behaved the same way in the car yet this is a common problem with many dogs. (It naturally depends on your circumstances. I know a vet in Dublin who, after several car break-ins, 'employs' his terrier as a car guard when he and his wife go out for the evening. Sweet as butterscotch anywhere else, this little dog turns into a sabre-toothed savage if anyone but her master approaches his car.) Nor would I be pleased if my dogs attacked the postman or meter reader or even implied that they might do so. It's easily possible to have too much of a good thing and this is true with protective aggression. Guarding can sometimes become so extreme that friends and neighbours no longer call.

Prevention

Both dominance aggression and fear aggression can be components of protective aggression. This means that proper selection, followed by good socializing, are equally important factors in preventing excessive protective or territorial behaviour. Certain breeds, specifically the herding and guarding breeds and most terriers, are genetically programmed to protect. If these dogs are not properly socialized, they are likely to become protective of their territories but in an undisciplined and uncontrolled fashion.

Maternal aggression is a hormonally influenced form of protective aggression in females but, although it can be one of the most ferocious forms of canine aggression, it is self-limiting and only lasts as long as the female is either pregnant or nursing. The simple way to avoid maternal aggression is to avoid confrontations when the dog is in pup or feeding her young.

Treatment of protective aggression

Females that exhibit protective aggression during their false pregnancies, protecting toys and slippers, can be treated by simply removing the coveted objects and either letting the hormonal cycle run its course or treating with appropriate hormones to counteract the hormone of pregnancy. Hormonal treatment is very much hit and miss simply because we know so little about specific hormonal effects but there are both tablets and injectible drugs licenced for use during false pregnancies.

Because there is almost always a considerable learned component to protective aggression, treatment is usually possible through operant learning and counter-conditioning. What complicates treatment is our usually ambivalent attitude towards canine protective aggression. Most of us actually appreciate a little protective aggression from our dogs. We enjoy the feeling of fidelity and loyalty we infer from our pet's behaviour. This means that there are several options in treatment.

The simplest treatment is to counter-condition the dog to be friendly with anyone who come on to his territory. This is done by withdrawing all rewards from the dog and only rewarding him when visitors arrive. All members of the family ignore the dog and give no attention or affection except when visitors come over. Sometimes you might even ask the visitors to feed the dog. When visitors are present, the dog gets as much affection and attention as you care to give him, as long as he shows no signs of aggression. Usually within a relatively short length of time he will become fully counter-conditioned and will look forward to visits from strangers.

In most cases we don't actually want our dogs to slobber over any stranger who comes into the home. All surveys consistently have shown that over 75 per cent of dog owners look upon their dogs for some form of protection. This means that treatment is best directed at desensitizing the dog to show no aggression under specific circumstances, such as when you are at home. This is done through voice control and discipline in the following way. As an example I'll use a dog that barks when the doorbell rings and then shows aggression when the stranger enters your home, but this treatment is equally applicable to dogs that, for example, try to attack the postman.

1. Control your dog by teaching or re-teaching the commands to 'come', 'sit', 'stay', and 'down-stay'. (see Appendix)
2. Temporarily withdraw affection and rewards, only rewarding obedience to your commands.
3. Train your dog to sit quietly some distance from the front door.
4. Once this has been achieved, command him to sit, then go to the front door and let him see you ring the bell. Reward him if he doesn't bark. Discipline him with a smack or a few minutes' banishment to an empty room if he barks and then ignore him afterwards before trying the routine again.
5. Once he shows no signs of protective aggression when you ring the bell, carry out the same procedure but open and close the door.
6. As he remains under your control, increase the intensity of the training by first having someone else who the dog knows ring the bell, then ring the bell and enter the hall and stand, then ring the bell, enter the hall, stand and leave, all the time rewarding the dog for sitting quietly under your control. If the dog fails at any stage, go back to the previously successful stage before ending the training episode.
7. Once he tolerates a familiar person ringing the bell and entering your home, switch to someone who is a stranger to the dog and repeat the series. If at any stage the dog shows aggression, isolate him and ignore him until you repeat the training at the previously successful level.

This is a long and laborious task and, of the various forms of aggression, protective aggression is one that is likely to respond to physical punishment. A dog trainer friend cured protective aggression in a Cocker spaniel this way. The dog had developed a classic conditioned response to the door bell. On hearing it, he experienced immediate pituitary-adrenal stimulation and would attack anyone who entered the hallway. Armed with a thick telephone book and the front door key, the trainer silently unlocked the door then rang the bell. The Cocker, as if shot from a cannon, burst towards the door where its head met the telephone book. The trainer immediately shut the door and went off for a cup of coffee returning an hour later when he repeated the procedure. The dog behaved in the same fashion. Half an hour later the same happened again. I should say that although telephone books can look quite punishing, the effect is more startling than painful. Fifteen minutes later the Cocker once more charged the door and met its surprise and it did so again ten minutes later. On the next ring, ten minutes after that, he simply peered around the corner from the kitchen and stared. In less than two hours he was counter-conditioned – desensitized to the doorbell and strangers. Punishment works on the dog's mind but it must be applied with care.

Reward based treatment of protective aggression can result in dogs loving intruders but sometimes this is the only effective way of containing the problem. We naturally like to have it both ways. We want our dogs to selectively welcome our friends and protect ourselves, our homes and our property from intruders. It's difficult to have it both ways however and a compromise is often necessary.

5. INTER-MALE AGGRESSION

Males of all domesticated species, including dogs and us, have a special propensity for fighting with each other. This type of aggression is quite rare in females. When two females fight, it's usually because of a conflict in dominance. The curious fact is that, although serious inter-male aggression only occurs after puberty in dogs, it is not wholly dependent on the male hormone testosterone. Giving testosterone to adult spayed females, for example, is relatively ineffective in making them more aggressive.

Males of most mammalian species have a propensity to fight with each other.

It appears that inter-male aggression has its origins in the neonatal development of the pup and is initially stimulated by the early masculinization of the pup's brain just before birth. The testosterone surge at puberty is, of course, the precipitating cause of this form of aggression but it is only influential on the already masculinized brain. Receptors must be present for the hormone to induce aggression.

Inter-male aggression is responsible for approximately one out of every ten cases of aggression treated by veterinarians and is, of course, a specific form of dominance aggression. It can be induced by the sight of another male dog or by the scent, the pheromones that are emitted.

Prevention

Although puppy testing has been shown to be of little value, the one facet of testing that reliably predicts future behaviour concerns juvenile aggression. Pups that act in a masculine dominant way are more likely to develop inter-male aggression problems when they mature. To avoid this type of problem avoid dominant male pups. And naturally, if you want to avoid this problem completely, choose a female as a companion rather than a male pup.

Treating inter-male aggression

This is the only form of aggression for which there is a good likelihood for improvement through castration. Clinical experience in Europe and North America has consistently shown that there is approximately a 60 per cent likelihood that castration will significantly reduce inter-male aggression. The loss of testosterone occurs within 24 hours of surgery. This means that the neural system in the brain, that was originally stimulated by the surge of testosterone near birth, no longer receives its hormonal messages. At the same time, the general odour of the dog changes and he is no longer as 'offensive' to other dogs. He still remains a male however. Every cell in his body is male and his brain remains masculinized. The difference is that the active hormonal influence on his mind has been removed.

When castration alone is not effective in reducing inter-male aggression, the female hormone progesterone can be used. This increases the success rate from 60 per cent to 75 per cent. Success can be further increased through counter-conditioning as in dominance aggression.

Because inter-male aggression and dominance aggression are often intertwined, there can be a problem in deciding who to treat and how he should be treated. As we know, dominance aggression between dogs is treated by expanding the social distance between two dogs. If their aggression with each other also contains a degree of inter-male aggression, castration is still the treatment of choice but it is the less aggressive dog, the more subservient one, that should first be castrated. This expands the natural social distance between them and eliminates the pheromones from the less dominant dog that provoke the more dominant one. Counter-conditioning as in simple dominance aggression problems should be carried out too. If these procedures don't have the desired effect over the following five to ten weeks, then the dominant dog should be castrated too. Once more, this should be done in conjunction with counter-conditioning for dominance aggression as I've previously described.

In veterinary practice I am frequently asked to castrate dogs that are aggressive. The facts are clear, however, that the operation is pointless in many, if not most, instances of aggression. Inter-male aggression is one, however, where the procedure, a simple operation, is highly likely to improve the behaviour of the dog.

There are many other forms of canine aggression but these affect the mind of the dog much less frequently than the ones I have mentioned. There are three that are worth mentioning.

PREDATORY AGGRESSION

Dogs are predators. Left to revert to natural behaviour, the domestic dog becomes the dingo. It doesn't matter how much we modify their genetic basis, it is a factor that cannot be eliminated. Selective breeding and training have, however, dramatically reduced predatory aggression in dogs, so much so that veterinarians are rarely consulted over this type

True predatory aggression is rare in dogs and usually only occurs in the form of stock chasing.

of behaviour. The basis for predatory aggression is genetic but it must also be learned from the mother. The naturalist David Mech has spent a lifetime observing wolves and has described how a wolf pack 'thinks' when it engages in group predatory aggression. Much of what he describes applies to playful dog behaviour but sometimes the play ceases to be play and changes to true predatory aggression.

Mech spent one winter observing a wolf pack hunting moose. In the summer, wolves will prey on almost anything, small mammals, birds, snakes, lizards, berries, fish, insects, earthworms and grass. In the winter, in Canada, their diet is restricted to large ungulates. While Mech watched, there were 120 'moose detections' and six successful kills, a 5 per cent success rate. Mech feels that a lot of the unsuccessful hunting is really simply a system of testing methods and honing principles and the same probably applies to predatory aggression in dogs. Mech never saw a moose kill when the moose stood to defend itself. Nor did he see a kill if the moose kept running. It simply outpaced the wolf pack. He only saw kills when moose were stalked, then taken down on the run.

We see this type of natural behaviour in most dogs when they play at prey catching. Dogs often assume their wolf posture, slinking up to attack. They capture after short chases. Long chases are rare and dogs give up easily. When dogs engage in predatory aggression for real, it is usually against different species, sheep, cats, squirrels, hedgehogs, groundhogs, but it can also be directed at children or older people. Although it is surprisingly uncommon, this is still the most serious type of aggression because the aim is to kill. In areas where there are large numbers of stray dogs, small packs can form and in rural areas, these can become predatory packs chasing down and killing sheep, goats and even cattle. It is a myth, however, that urban stray dogs engage in predatory aggression.

Thomas Daniels and Alan Beck have each studied free ranging domestic dogs in Newark, New Jersey and Baltimore, Maryland respectively. These dogs lived in densities from 400 to 600 per square mile, far greater than the living densities for any wild canids such as coyotes or wolves. Both observers noted that aggression of any kind was rare and that free ranging dogs rarely formed packs. Ian Dunbar came to the same conclusions when he studied free ranging dogs in Berkeley, California. In all of these studies, the dogs were roaming because of curiosity, habit or boredom, rather than hunger. Instances of predatory aggression were insignificant. When it does occur, however, it is an overwhelming problem.

Prevention

Any breed of dog can behave in a predatory fashion, but a dog is less likely to do so with animals to which it has been socialized. Raising a dog with rabbits and cats reduces the likelihood that it will ever attack rabbits and cats. If you live in a rural area and your dog will be allowed to roam over your land, it is best to expose him during his socialization period to the various species he will later come in contact with. This is what shepherds do with their sheep dogs. The dogs are raised with sheep. Then, when they are older, their natural predatory inclinations are channelled into sheep gathering rather than sheep killing. Some guarding breeds such as the Maremma in Hungary were historically raised with sheep and even bred to look like them. Their natural inclination towards predatory aggression was overcome through early socialization and their protective aggression used to protect their flock.

Treating predatory aggression

This is the most difficult form of aggressive behaviour to overcome, simply because it is rooted so deeply in the dog's mind. The rewards are, after all, magnificent. The chase and the kill. This transcends all the modifications we have ever imposed on the dog and strikes right through to his very core. Counter-conditioning is often pointless because it is so difficult to find rewards that are anywhere near as vital as the reward of the chase. The only possible treatment of the condition, other than confinement, is through severe punishment. This is one form of aggression where electric shock collars, used under professional supervision, might be the only treatment. The only alternative is euthanasia and remember, euthanasia for behavioural problems is the most common cause of death in young dogs.

IDIOPATHIC AGGRESSION

Sometimes some dogs become aggressive towards people they know and for no apparent reason. The aggression isn't related to dominance, fear, rivalry, jealousy or any other known cause, and in these circumstances it's called idiopathic aggression – cause unknown. The cause is often there, however, but it's hidden deep in the dog's genes.

Certain breeds suffer from idiopathic aggression. The Pyranean Mountain Dog used to and the Bernese Mountain Dog does today. In Holland, the problem in Bernese Mountain Dogs was traced back to two males that had been imported into that country. Other breeds can suffer from this behavioural problem, St. Bernards, Rottweilers, Dobermanns, German shepherds and most frequently, solid coloured Cocker Spaniels, especially blond English Cocker spaniels. In this breed, the condition is known by two colloquial names, the Jekyll-Hyde Syndrome or Avalanche of Rage Syndrome. Typically dogs that suffer from this type of idiopathic aggression are usually affectionate, obedient, pleasant, well-mannered canines who, for no reason whatsoever, suddenly and ferociously turn on their owners or visitors, biting legs, arms or faces. The observant owner might notice a glazed other worldly look to their dog's eyes just before an attack but other than that, there is no warning. Some dogs will remain subdued for a short while afterwards. Others snap out of it almost immediately and return to their normal affectionate ways.

Clinical examination of these dogs rarely reveals any abnormalities. They have to be there but the problem is buried deep in the dog's brain and is almost undoubtably genetic in origin.

Prevention

If the incidence is high in a certain breed or a certain colour within a breed, be extra careful in selecting a pup. Make sure that the problem does not occur in his immediate family. Check back three or four generations in breeds like Rottweilers, golden Cocker spaniels and Bernese Mountain dogs before choosing a pup.

Treating idiopathic aggession

There is no treatment for idiopathic aggression. Because it may clinically be related to certain forms of epilepsy, the use of anticonvulsant drugs might be effective, but the only treatment for this type of aggression is euthanasia. That's a harsh verdict but one that I have become resigned to after seeing the problems that ensue when people try to live with the condition in their dogs.

Before looking at other developments of the dog's mind, there is one final form of aggression that I should briefly mention and that is learned aggression.

LEARNED AGGRESSION

It is, of course, self-evident but dogs can be taught to be aggressive. People who train dogs for fights choose dogs from breeds that were previously selectively bred for fighting, breeds like bull terriers, and then often use pain to induce aggression. Pain is a powerful stimulus for aggression. More positively, most forms of dominance and protective aggression can be enhanced or redirected to our advantage. This is the basis for the use of dogs in crowd control. The advantage of learned aggression is that in the right hands, it can be turned on and off. A police dog, for example, must at one moment walk down a crowded street and be possibly petted by a child without any risk to that child, yet might in the next moment be required to attack a criminal. In the best circumstances, learned aggression involves retrieving as much or more than it actually involves aggression. In training, the dog learns to retrieve. First he retrieves a padded piece of wood. Later on, that becomes the villain's arm. At the same time, he might be trained to bark or growl, to be as intimidating as possible. Police forces with the most professional canine corps use this form of learned aggression and the dog's mind is eagerly open to it. Unfortunately, the same cannot be said for guard dogs, many of whom are treated inhumanely or even painfully to make them aggressive.

Aggression develops to its maximum potential by the time a dog is two and a half years old. We have an ambivalent attitude towards it. On the one hand, we dislike unruly, aggressive dogs. On the other, we admire the dignified stolid, well-mannered canine who eventually turns and uses aggression on the unruly one. We expect our dogs to let our friends enter our homes and to cheerfully greet them but, at the same time, to prevent intruders from entering and to deal with them severely if they do so. We fantasize that our dogs will protect us from bears and tigers, yet expect them to leave sheep and rabbits to graze freely. We ask too much of the dog's mind. It is as plastic and malleable as that of any other domestic species but, in the field of aggressive behaviour, there have to be compromises between our needs and their abilities.

Chapter Nine

Social Behaviour – Eating, Exploring, Eliminating

The rules of canine learning will continue to apply throughout the dog's life and will affect all aspects of his social behaviour. But, once more, remember, the overwhelming difference between the dog and all other members of the canine family is that we humans make up the other members of the dog's pack. All of his behaviours, the way he eats and eliminates, the comfort that he seeks, the way he explores, his fears and phobias, even the way he has sex, all of these behaviours are influenced by us. We have a dramatic effect on the dog's mind, far greater than has been previously recognized.

EATING AND EXPLORING BEHAVIOUR

As a general rule, most carnivores spend a great deal of time hunting and very little time eating. David Mech quantified this when he watched his wolf pack and saw that only 5 per cent of chases resulted in kills. Because we feed our dogs, they don't have the opportunity to behave as they have been genetically programmed to do. The consequence is that natural hunting behaviours are channelled into other areas that benefit the human pack. Michael Fox has listed natural canine hunting and eating behaviours and I have modified his list to include the following:

1. *Stalking and pointing*
   This has been modified for play activity in some breeds or for shepherding and hunting, especially in Collies and Pointers.
2. *Herding*
   This hunting behaviour has been successfully adapted in breeds around the world to herd sheep and cattle.
3. *Digging (when prey goes to ground)*
   All dogs will dig but some will do so more than others. Terriers bred for rabbiting and foxing are notorious diggers but because food

hoarding is a natural behaviour in dogs, all have the prewired tendency to dig.

4. *Scent following*

   Dogs hunt by following ground and air scent. Tracker dogs use this hunting ability to follow human scent.

5. *Shaking, throwing, catching, killing*

   In most dogs this has been modified to a play activity. Dogs will shake their toys and catch balls. Some terrier breeds still have a strong genetic potential to shake and kill smaller mammals.

Retrieving is simply a variation of the wolf's prey carrying activity.

6. *Bringing food back to the den (retrieving, food carrying and 'gift giving')*

   This natural canine behaviour is most apparent in the breeds bred to retrieve game but can occur in any breed. Terriers as well have a strong genetic predisposition to carry objects in their mouths although they are less likely to relinquish them than are retrievers.

7. *Regurgitating*

   This is a behaviour that has been considerably bred down in the dog compared to the wolf. Dogs still do, however, have a sensitive vomiting reflex and both males and females are still known to regurgitate food.

8. *Gnawing and chewing*

   This is a comfort behaviour, carried out more frequently by dogs than by wolves simply because we deny dogs much of their natural chewing.

9. *Eating grass*

Dogs are omnivores. Grass eating is normal and I will discuss this shortly.

10. *Prey-carrying*

Some dogs develop the habit of taking their food out of their food bowl, carrying it to another place and eating it there. This is based on the wolf's prey-carrying activity.

11. *Prey-guarding*

Some dogs will guard their food. Some animal behaviourists say that this activity is intimately associated with dominance aggression.

12. *Prey-burying* (*food cache*)

All dogs bury bones but few ever dig them up.

Stalking has been modified for play activity in some breeds and for shepherding in others.

All of these behaviours are carried out in conjunction with the rest of the pack and are perpetuated in dogs to varying degrees, often for our benefit. Selective breeding, for example, has intensified stalking, pointing, herding, retrieving and digging behaviours in specific breeds. This is marvellous for the dog and marvellous for us. The dog benefits from activity and we benefit from their inate abilities which are superior to ours. It is the rare dog today, however, that is allowed to hunt even in this modified form. The consequence is boredom, one of the most overwhelming factors in the life of most pet dogs. Most of our pets lead restricted dull lives where the only excitement is the return of members of the family, a little exercise and eating. One of the consequences is that eating problems that would never occur in the wild are common in pet dogs. Another is that roaming, the pet dog's answer to boredom, is responsible for the free range dog problems that afflict most major cities in Europe and the Americas. Let's look first at eating behaviour.

Dogs wolf down their food. This is a genetically programmed behaviour. To the dog's mind, it's a matter of feast or famine, and his behaviour reflects the fact that, after a kill, there is competition amongst the pack for the prey. The faster you eat, the more you get and there might not be another kill for several days, or even weeks. The animal behaviourist Roger Mugford has noted that a Labrador can consume 10 per cent of his body weight in one meal. The fact that the Labrador is a house dog and that there is in fact no competition for the meal is irrelevant as the behaviour is genetically programmed.

Feeding dogs in groups is almost guaranteed to increase their individual food intake and this is a common ploy used by veterinarians to make sure hospitalized dogs eat.

Hunting and eating in carnivores involves dominance and aggression, as we have already discussed, with the problem of predatory aggression, but through selective breeding for over a thousand generations, natural predatory aggression in dogs has been channelled into stalking, pointing, herding and digging. These behaviours, together with shaking, carrying, guarding and burying, all manifest themselves in what can be called biologically dead end behaviours in domestic dogs.

FOOD

Although we usually think of dogs as carnivores, they are in fact omnivores. Dogs will eat grass, berries and roots as well as meat. And like squirrels, they hoard food for future consumption. Most wild canids will do this. The ethologist Nikko Tinbergen once watched a red fox steal gull eggs and bury them in sand on the beach, digging the hole with his forepaws and covering the eggs by pushing sand with his nose. Later that year, after the gulls had left, the fox returned and one by one, dug up the eggs and ate them.

Dogs will bury food too but they rarely consummate the behaviour by retrieving it and eating it afterwards. All breeds can do so. The Yorkshire terriers I grew up with used to half bury their bone shaped biscuits in a large indoor garden bed, in our home, making it look like some miniature war cemetery. In the absence of any earth, my parents' present dog pushes biscuits into every corner of every room in their home and will nose the carpet for fifteen minutes trying to 'bury them'. And in practice, I meet one rather corpulent Labrador that opens his food cupboard, takes out a can of dog food and buries it in his back garden. His owners keep a shovel by their back door to dig up his evening meal.

Selective breeding enhanced natural retrieving behaviour in dogs.

The owners have to do so because, although food caching is still an integral part of the dog's mind, retrieving it is not.

Eating grass is another normal canine behaviour that surprises many dog owners. When a wild canid, a fox or wolf for example, captures prey, it eats all of it including the skin and the intestinal contents. Most of the prey are herbivores and the intestines are filled with roughage. Some dogs will eat grass to apparently make them vomit but others will graze on long grass for the apparent sheer pleasure of eating it.

Dogs explore with their mouths. Young dogs in particular, when coming upon something new or unusual, will both smell and taste it. This behaviour sometimes becomes exaggerated and they eat the object and if it is undigestible, pebbles for example, it can lead to intestinal obstruction. The problem has a human factor because more often than not, the actual swallowing results from the owner being unable to command the dog to obey. When the owner tries to recover the undesirable object, the dog runs off and swallows it. This might happen for two reasons. First, the owner might not have enough dominance over the dog to recover the object. And second, the 'chase' actually rewards the dog. In his mind, he senses cause and effect. If he picks up a pebble, or a worm, or more commonly his own or another dog's faeces, he learns that he gets a response from his owner. It is a simple case of operant learning where the dog gets a hidden reward, attention from his leader, when he behaves in a certain way.

Eating faeces, or coprophagy, can be 'normal' in the sense that it is a normal behaviour of females to eat the droppings of their pups and it is also a normal behaviour of dogs to have an instinctive preference for decaying food, for rotting carcasses, rubbish and faeces. There is, however, no uniform explanation for the habit. Some veterinarians argue that coprophagy is a sign of dietary imbalance and argue that a diet high in carbohydrates and starch contributes to the problem. Similarly, not enough food or a very low protein diet might stimulate the behaviour. Even the consistency of the stool, hard or frozen, seems to contribute to stool eating. Others say that it can be an attention getting behaviour as I've just described. The answer is probably that coprophagy is self-rewarding in many ways. First of all, to the dog it tastes good. Finding the faeces involves exploring and that stimulates the mind. And if it happens to be horse droppings or sheep or rabbit or cattle stools, it's actually quite nutritious. In most instances dogs seem to grow out of the habit, but if you want to treat the condition do the following:

1. Feed a consistent and nutritious diet.
2. Feed enough food twice a day to meet all the dog's requirements, making sure that the diet contains good levels of fat and protein.
3. Prevent the dog from eating stools by either avoiding areas where it lies or by muzzling him.
4. Stay with the dog and distract him when he sniffs any stool.
5. Discipline him if he eats any.
6. In extreme circumstances, 'flavour' any deposited stools with red pepper to act as an aversion to eating.
7. Train the dog to defecate in a specific area and always keep this area clean.

The possibility that a dog might eat faeces because of an inherent dietary imbalance is fascinating and has broad implications. It is known that specific species of animals have specific hungers for certain vitamins, minerals and amino acids and will alter their diets to make sure they take these substances in. Classic experiments were conducted almost 50 years ago showing that when rats were offered a 'cafeteria' of food, they showed considerable nutritional wisdom in selecting and consuming a nutritionally balanced diet. This might be true for rats but is not so for domestic dogs. Dogs will not wisely choose what is good for them. They will choose food based on their early learning. Ake Hedhammar at the Swedish University of Agricultural Sciences in Uppsala noted in the early 1970s that growing domestic dogs fed ad lib, simply overate and played less. Early learning teaches for life but there are several unexpected factors that also affect the dog's mind.

Obesity

It's extremely rare to see fat canines of any species in the wild and this is because most animals have a refined ability to adjust their consumption of food to meet their energy requirements. In hot weather they eat less. In cold weather, when they have higher energy requirements, they eat more. The control of food intake is still poorly understood but there are chemical factors acting on the brain that are responsible. In one experiment, blood from a hungry and from a full sheep were cross-circulated.

Obesity in dogs is a man-made problem. It is almost unheard of in any canine species in the wild.

When this was done, the full sheep became hungry and the hungry sheep became full. Hormones released from the intestines, together with glucose and various amino acids all act on receptors in the hypothalamus to give the dog a feeling of a full stomach. This is how most animals maintain a normal body weight without apparently working at it. Nevertheless, obesity remains a considerable problem in dogs. Most estimates in Europe and North America suggest that one out of every three dogs is fat. How can this be when they are so finely biologically tuned to not get fat. The answer, of course, lies with us.

To begin with, we have allowed dogs that would have been too fat to survive in the wild, to survive and reproduce. There is now a genetic predisposition to fatness in certain breeds rather than in others, Labradors and Spaniels for example.

Evidence from studies in rats suggests that the brain opiates, the endorphins, might exist in higher levels in overweight animals. Perverse as it sounds, the housebound dog leads a stressed existence. It cannot express its natural behaviours, stalking, chasing, exploring, investigating. The brain endorphins might increase in states of anxiety and be in some way related to obesity.

Another controversial reason for obesity in dogs concerns how fat they are allowed to be as pups. Some experts argue that the number of fat cells that a dog has is genetically set and that throughout its life, the dog's mind will 'defend' this number of fat cells. The argument goes that if a pup is fat at or soon after birth then it will always remain fat because this is the metabolic state that the brain 'defends'. Supplementing the mother's milk and over feeding the pup as an infant could mean that weight loss later in life would be almost impossible, akin to starving the dog.

Regardless of the specific causes, most cases of obesity in dogs are owner inspired but there is one hormonal cause of obesity that is both important and controversial and that is the influence of neutering on weight.

Sex hormones and weight

Male dogs eat more than female dogs. Male dogs, perhaps surprisingly, also carry more of their body weight as fat than do female dogs. The natural conclusion then, is that sex hormones affect eating habits. We already know that this is the case. The female hormone estrogen suppresses the appetite and females experience sometimes dramatic changes in their body weight, eating behaviour and general activity in relation to their reproductive cycle. As I write this, my younger retriever is just finishing her estrus cycle, has been eating 25 per cent less and walking around with a rain cloud over her head for three weeks.

But what effect, if any, does neutering have on the dog's appetite and on obesity. Experimental evidence in rats conclusively shows that neutering results in increased food intake and changes in energy balance that together will result in weight increase and obesity, but research in dogs is more equivocal. First of all, dogs cycle twice yearly compared to rats that cycle every four to five days. Katherine Houpt at Cornell State University carried out the most detailed experiments on the effect of neutering on body weight and reported that spaying females resulted in a gain in body weight and food intake after surgery. She reported only a two to three pound weight increase and said there was no increase in subcutaneous fat deposits, but concluded that spaying will lead to slightly increased body weight. I have found that with my own dogs, I have had to decrease their calorie intake by about 25 per cent after spaying in order to maintain them at their previous weight. I also know that in the case of my own dogs, there has not been a reduction in activity or exercise after surgery. These are the circumstances I see with client's pets as well. Many dogs, after spaying, continue to have the same amount of exercise but become obese on the same number of calories that they consumed before surgery. Dogs do not have to become fat after spaying, but there are undoubted metabolic changes that occur when the ovaries are removed. Standard advice then is that when dogs are spayed, monitor their weight carefully after surgery and at the first sign of weight increase, reduce their calorie intake initially by 10–20 per cent. Spayed dogs do not become lethargic but fat dogs do.

Anorexia

Dogs may occasionally become reluctant to eat for other than medical reasons. Large breeds of dogs are less selective about their diet and this is only natural. From a genetic point of view, large breeds should be programmed to eat anything that is nutritious. This is less important in smaller breeds which is one of the reasons why apparent anorexia occurs in breeds like Yorkshire terriers more frequently than in German shepherds.

Anorexia can sometimes be, in fact, a food allergy, masquerading as a food aversion. Some dogs are allergic to the gluten proteins found in wheat and feel distressed after eating foods that contain these proteins. They develop a generalized aversion to commercial foods and we think they have no appetite when in fact they are being quite sensible.

Dogs can suffer from true anorexia when there is an abrupt change to their routine, kennelling, a new home or the loss of a person or another dog. It seems similar to depression or mourning in people. In most instances, the dog will regain its appetite but it's often necessary to feed tasty foods or even to hand feed. It's important to remember that dogs are always learning, even in situations such as these. If anorexia continues, and medical or physical reasons for the condition have been eliminated, it has to be considered that the dog is using anorexia as an attention getting behaviour. Especially in circumstances where there has been an abrupt change in the dog's circumstances, the loss of a human or canine companion, or a new home, the security of constant attention from someone is very gratifying and some dogs will rapidly learn that they can perpetuate this attention by only eating when hand fed. In other words, we reward the anorexia with attention. Curing the problem simply involves ignoring the anorexia and rewarding good eating. I have never seen a dog starve to death or heard a true story of a healthy dog starving to death when food was available.

ELIMINATION BEHAVIOUR

Adult wolves are fastidious about their sanitary habits and empty their bowels twice daily away from the den. This type of behaviour is pre-wired into their minds. In the distant past, the animals that were most fastidious in their sanitary habits were more likely to raise offspring that didn't suffer from heavy loads of intestinal parasites. These were the evolutionary survivors.

Urine is left at nose level as a territorial scent marker.

The communication aspects of elimination and the classical conditioning of the behaviour have already been discussed, but there are a few other points that should be mentioned. Male dogs don't lift their legs to urinate until about six months of age. The lift improves with age and reaches its highest point by about two years of age. Females usually squat although some do lift a leg slightly and it's thought that these females might have been mildly masculinized by testosterone while still in the uterus.

Fastidious as they are with their body waste, dogs still cause their owners considerable problems with their elimination behaviour. In fact, one out of every four or five behaviour problems presented to the veterinarian concerns an elimination behaviour. Dogs might urinate in the 'wrong' place because of:

1. lack of house training
2. territory marking
3. anxiety over being separated from the family
4. fear
5. excitement
6. oversubmission
7. wrong diet
8. to get attention
9. disease
10. inbreeding

11. poor early experience

Before tackling the problem of a dog urinating indoors, you must first determine exactly why he is behaving in this way. This means establishing general facts about his behaviour.

1. Lack of house training

I have already discussed housetraining under classical conditioning but there are these points to remember.

(a) Don't leave the dog loose in your home and unsupervized until he is housetrained

(b) Take the dog outdoors when he wakes, after he eats and after exercise and play

(c) Confine him to a small space when he is not under your supervision

(d) Praise him whenever he urinates in the proper place. Punish him ONLY if you catch him urinating in the wrong place

(e) eliminate odour from soiled areas using alcohol or any effective proprietary substance. (Remember that it's almost impossible to mask urine odour on concrete.)

2. Territory Marking

Territory marking is testosterone induced. It's a male dog's behaviour although females will occasionally urine mark as well. Urine marking reduces a dog's anxiety by masking any other dog's odour with his own. Young dominant dogs around two to three years old urine mark more frequently than other dogs and are more likely to do so when they feel threatened, such as before a fight or when a new dog enters the environment. Any type of emotional stress can induce urine marking, new people, female dogs in heat, separation anxiety, the chance of mating or the visit to the veterinarian.

Urine marking indoors can be dramatically diminished through castration. The behaviour is usually reduced by 50 per cent or more, although outdoor marking often continues at its previous level. Spaying females will also reduce urine marking if they have done so during estrus. The use of a synthetic progesterone drug will also dramatically reduce the behaviour in dogs. Sometimes all that is needed is a two week course and the effect can be lasting. If these treatments are not possible, punishment and counter-conditioning will work but only if you actually catch the dog in the act.

3. Separation Anxiety

Separation anxiety usually causes house soiling rather than urinating and I will discuss this, together with barking and destructive behaviour, in the next chapter.

4. Fear
Urination and defecation is a common response to acute fear in some dogs. These dogs often empty their anal glands at the same time. The problem occurs when they are placed in a situation from which they can't escape. Treatment quite obviously involves identifying what is so fearful and either eliminating it or reducing its intensity.

5. Excitement
Urinating because of excitement usually occurs in young excitable pups, often during greetings or while playing. This can be quite different from either submissive or fear induced urination and, to eliminate the problem, either train the dog to be more quiet, make sure he empties his bladder before play or simply avoid exciting situations for him.

6. Oversubmission
Very submissive dogs will often urinate when they are simply looked at, let alone touched. In these circumstances, avoid reaching down and stroking the dog from above and avoid intimidating eye to eye contact. Treatment involves enhancing the dog's self esteem. Don't greet it when you come in but rather, let it settle and relax. Open the door and let it outside to urinate. Only when it is settled should you talk to it and this is best done by getting down to its level and stroking it under the chin, a less intimidating position.

7. Wrong Diet
Excessive urination is extremely rare but possible especially if the diet is sodium rich. House soiling is more likely, particularly if a diet results in increased stool quantity or changed consistency.

8. To Get Attention
Under emotional stress, some dogs can drink excessive amounts of water. This is called psychogenic polydypsia and is a genuine psychosomatic illness. I've only seen it once in a Cavalier King Charles Spaniel. This dog drank massive quantities of water, three or four litres a day for three or four days each time he returned home from kennels. The problem was self-limiting in that his behaviour always returned to normal. If it were not to do so, the treatment would simply be to permit only necessary drinking.

9. Disease

Many diseases can cause excess drinking (polydypsia) and excess urinating (polyuria). Different forms of diabetes, liver diseases, kidney and bladder diseases and central nervous system problems can all lead to excessive urination. Treating these causes of excessive or inappropriate urination naturally involves curing or controlling the disease itself.

10. Inbreeding

There are certain breeds that just seem less 'houseproud' than others. While breeds such as Labradors and German shepherds are almost invariable easy to house train, others like Yorkshire terriers, Maltese terriers, and Bichon Frises are conundra. The increased difficulty in housetraining these dogs is probably genetic in origin and treatment is consequently difficult. It simply means that you must persevere longer and work harder to train these dogs to eliminate in the appropriate places.

11. Early Experience

Scott and Fuller's original work on the genetics of dog behaviour revealed that pups that had been kept in cages until three to five months of age and that were deprived of the opportunity to urinate found it difficult to impossible to ever become house trained. While at university, I 'rescued' a laboratory Beagle pup and found to my regret that this was the case. There is no effective treatment for this lack of early learning.

Exploring, eating and eliminating behaviours all develop to maturity very early in the dog's life. The dog's natural inclination is to be inquisitive, to eat almost anything, to eat it rapidly and to keep his den clean. These are central reasons why we chose dogs to work for us and to live with us in the first place. They have many other social behaviours, attachment and comfort seeking are important ones. When these are threatened the dog can behave in an excited or agitated way and that is what I will discuss next.

Chapter Ten

Social Behaviour – Fears, Phobias, Anxiety, Excitement

The dog has become our closest companion because we share so many behaviours with him. We often misinterpret some of his behaviours. We anthropomorphise his actions and, for example, impute guilt when a dog has been destructive when in fact his response is more likely to be fear. In circumstances of anxiety, stress, fear, phobia and excitement, however, the underlying psychological experiences in the dog's mind are probably acutely similar to those that we experience when we have these feelings. The physiological changes certainly are.

FEAR

Fear in a potentially harmful situation is a normal and healthy reaction. We humans have the distinct advantage over all other animals in that we can talk about it. Imagine what a thunderstorm would be like if no one could tell you it wouldn't hurt us. The only way we would learn that thunderstorms aren't dangerous would be through habituation, through learning from experience that the noise would not harm us. This is the way dogs must learn to control their fears.

Fear in the dog's mind has many origins. Allegheny Sue's fear was genetic in origin. Her predisposition to fearfulness was inherited and this genetically 'pre-wired' fear behaviour is resistant to change. Inherited fear can take many forms. It might affect only one pup in the litter and is easily detectable at six to ten weeks of age. This pup will be wild eyed and withdrawn. Studies in rats have shown that a protein deficiency around birth can produce increased emotionality later on in life, and if this applies to dogs, it means that pups that start life as runts might be more likely to develop nervous or excitable personalities.

Inherited fear is sometimes seen in entire litters as was the case with Allegheny Sue's litter of Pointers. In these circumstances, exaggerated fear develops between three and ten months of age. These pups have a generalized fear of anything new or unusual, a global fear and once more, this behaviour is 'pre-wired' into the mind of the dog and is resistant to change.

Every dog, just as every human, will inherit a certain level of reactivity to his environment. This is normal and healthy. It contributes to our and their survival, but in some breeds or individuals, this inherited reactivity appears to be exaggerated. This is the case in certain German shepherds and other breeds such as pointers and is a third form of inherited fear response.

We already know that early experience has a profound effect on the dog's subsequent behaviour and nowhere more so than in his fear response. Scott & Fuller at Bar Harbour, Maine and Willson in Sweden have succinctly described how nervous mothers can imprint nervousness into their offspring, unwittingly teaching barking and excitable behaviour. We also know that an impoverished physical or social environment early in life will have an overwhelming influence on the emotional development of the dog's mind. Pups that are deprived of normal exposure to common stimuli during their critical periods of development quite simply become fearful dogs when they mature. There is considerable individual and breed variation but virtually all dogs will show a fear response to new and unusual stimuli that they have not experienced before they were sixteen weeks old.

The principles of learning, both classical and instrument learning, apply to the formation of the fear response, its maintenance, any increase in its intensity and its generalization to other situations. And finally, although there is no direct evidence that anxiety or fear in the mind of the owner can be transferred to the mind of the dog, this is an anecdotal suggestion that has a sound basis in my experience.

Dogs express their fear in many ways. Fight or flight, usually flight, is a common response to fear but dogs often can't flee from what frightens them. Instead they might bark anxiously and assume the body posture of fear, in which the tail is tucked between the legs, the back is arched,

The lips are retracted and the ears flattened when a dog exhibits fear.

the lips are drawn back, hackles are raised and the ears are flattened against the head. At the same time, there is autonomic nervous activity. The heart rate increases, pupils dilate and trembling might occur. The dog might even urinate and, of course, he might try to bite.

Fears are constantly being learned and unlearned. A dog might, for example, learn to fear young children because a child has pulled his hair. It couples the unpleasant experience with the child and avoids the child. But as the child grows and starts dropping food from his highchair, the dog starts to associate pleasant experiences with the child. The presence of the child means scraps of food and, through desensitization, he loses his fear of the infant.

PHOBIAS

When fears are not unlearned in this way, they become phobias. Dogs can develop phobias to loud noises, to traffic and, I'm afraid to say, unfortunately, to me. Phobias to veterinarians and veterinary clinics are probably just about as common in dogs as phobias to dentists and dental clinics are to their owners. This might happen because the first association was exceedingly unpleasant. This can certainly be the case with veterinary clinics where the first visit usually occurs when the pup is in the fear imprint state of emotional development and also usually involves physical manipulation, a thermometer in the rectum, bright lights shone directly in the eyes, the injection of a vaccine and sometimes the drawing of a blood sample. For older dogs that have never visited the veterinary clinic, it can be equally fearful because of the presence of other dogs and cats, the new odours, the sounds and to the dog's mind, the aggressive behaviour of the veterinarian. The fear becomes a phobia either because the initial experience was so unpleasant or because that first fear reaction itself, the increased heart rate, trembling and agitation, was so unpleasant. Fear does not, however, always develop into a phobia. It doesn't necessarily involve fight or flight but rather manifests itself through anxious behaviour and excitement.

EXCITEMENT

At the veterinary clinic, a dog will often show fear through heightened activity, by behaving in an excited manner. Physiologically speaking, his body's response is the same as in fear, trembling, increased heart rate and panting, but rather than fighting or fleeing, he becomes restless and extra alert. He might bark, show increased attachment to his owner by nudging and making physical contact. He might even urinate or defecate or carry out a displacement activity such as chewing on his lead or yawning. All of these behaviours are signs of high arousal of his adreno-pituitary axis, of his aroused state of mind.

We all know that some dogs are 'high strung', that their excitability is inherited, but it is also simple to unwittingly train a dog to behave in this way. Dogs that go nuts when you reach for their lead or that become uncontrollably excited when they get in the car are more likely to have learned this behaviour than to have inherited it. In these circumstances, their excitement, their state of arousal is not related to fear but is rather a sign of anxiety, a display of stress related behaviour.

ANXIETY

Dogs are motivated by various genetically predetermined drives. Hunger, thirst, sex, aggression and territory marking are all natural drives but most importantly, from the viewpoint of our relationship with dogs, so too is attachment. If we deprive a dog of the opportunity to carry out his normal behaviour, this can create a state of anxiety. Restricting a dog to a life of luxury in a centrally heated home where he is fed like a king and allowed to sleep where he wants might sound like a recipe for canine contentment, but it still carries the seeds for anxiety for two main reasons. First of all, he is confined and this is 'unnatural' for dogs. And second, he has to live with us and, although we are in many ways marvellous dog substitutes, we still unwittingly create conflicts in his mind.

Conflict in the dog's mind creates anxiety

Anxiety can be created by the most innocuous of means. Dogs imprint on to their families as pups and then forever have a feeling of attachment to them. This is what we interpret as fidelity and treasure so much from our pets. But let's say that an owner disciplines his dog by smacking him. The conflict is obvious. The dog is attached to his owner, permanently attached by early bonding, but is fearful of him because he gets smacked. The result is anxiety, a state of conflict which the dog shows by acting tentative or even barking at the approach of his owner. Dogs can exhibit anxiety by showing greeting signs, tail wagging and approaching while at the same time showing fear with the head held low. They might roll over and urinate, signs of submission. Rarely they might bite at the same time. I once rescued a German shepherd that behaved this way. She was raised in a family and imprinted firmly to people but after being sold, became desperately anxious about people because of beatings, fearful because she was kept isolated as a guard dog in a store room and excitable because there was no activity in her life. On greeting me, she would approach in a friendly manner and then attempt to bite if she saw my hand, a classic case of emotional conflict. (Rehabilitation was possible because her behaviour was learned and therefore treatable through counter-conditioning.)

Conflict doesn't have to be as dramatic as this. If a dog is treated inconsistently by members of the family, if one person lets him up on furniture and another reprimands him for getting on furniture, the conflict is obvious. Fortunately many dogs are exquisite and constant observers of human nature and learn the subtle clues we give that tell them when a certain behaviour is permissible and when it is not.

Excitement is not the only outlet for a dog's anxiety. They can also indulge in what are called displacement activities, activities such as chewing through doors, stripping wallpaper and digging through the carpet. These are the displacement activities of the confined dogs. Anxious dogs might also behave in what can only be described as a neurotic fashion. They might chase their tails or physically act in any other strange but stereotyped manner. Destructive behaviour is the most common reason that pet owners seek out veterinary advice for behavioural problems. This destructive behaviour can frequently be related to the anxiety of confinement but it can also occur in other situations. A medical condition commonly called 'lick dermatitis' is thought to sometimes be a displacement activity, in which the dog licks his forelimbs excessively when he is in a state of emotional conflict. The question that then arises is why some dogs react to certain situations with fear and phobias while other dogs react with excitement and activity. The answer undoubtedly lies in differences in reactivity of the nervous system.

We know, for example, that inheritance is a key factor but so too is early learning. We know that pups raised in an emotionally or socially impoverished environment are more likely to grow up to become anxious high strung dogs. But they can show this behaviour by becoming highly excitable or withdrawn. Pavlov described these states as excitation and 'inhibition'. The British clinical psychologist Valerie O'Farrell prefers to use Hans Eysenck's terms 'introversion' and 'extraversion'. Extravert dogs, says O'Farrell, are sociable and impulsive. Introvert dogs are more thoughtful and withdrawn. Eysenck says that neurotic extraverts exhibit their neurosis through being self-destructive and anti-social. Applying this to canines, dogs that mutilate themselves, destroy property and bark excessively are neurotic extraverts. The excessively shy dog, the fearful one, the obsessional dog would be described as a neurotic introvert. Because fear, phobias and excitement appear to have similar physical and psychological components in humans and dogs, it seems possible, even sensible, that this is one area where terms in human psychology can genuinely be applied to the dog's mind.

Prevention of fears, phobias and excitement

We have already discussed the genetic and early learning components of neurotic behaviour in dogs and know that the best prevention is to avoid breeding from dogs that are known to carry the genetic predisposition to produce fearful, excitable pups, and to gently expose pups during the early critical periods of their lives to as much sensory stimulation as possible. Behavioural problems will still occur however, separation anxiety, destructiveness, phobias, excess barking, excitement when you or visitors come home, excitement in the car and even fearful urination. All of these can have a learned component to them and, the larger the learned component,the more likely it will be that they can be successfully treated through counter-conditioning.

SEPARATION ANXIETY

Separation anxiety has no age, sex or breed predilection. Any dog can suffer from it although Elizabeth McCrave at the University of Pennsylvania says that dogs with separation anxiety are most likely to be mixed breeds from animal shelters. In her studies, she found that dogs with separation anxiety problems greet their owners excessively, jump up on them, pester and follow their owners throughout the day and generally stick like glue to them. Interestingly, she found that letting your dog sleep on your bed, feeding him from the table, taking him on errands or letting him up on furniture had no effect on separation anxiety.

This problem occurs most frequently when there is a strong attachment encouraged by one person in the family. It then becomes serious if that attachment becomes threatened by a sudden change in routine. Separation anxiety occurs in dependent dogs. They might be on one hand territorially protective but on the other, they are dependent upon their master, following him around, constantly demanding attention and greeting him in the most maudlin and excessive way even after the shortest of departures.

These dogs can show their distress before the owner leaves either through excitement or depression but the most important signs of separation anxiety occur within half an hour of the owner's departure. These include:

1. Aggression when the owner leaves – growling or nipping at ankles
2. Destructive behaviour – chewing and digging and tearing
3. Self multilation – excessive licking
4. Hyperactivity – constant pacing, stereotyped behaviour such as running up and down the stairs
5. Urination or defecation
6. Psychosomatic problems – diarrhoea, vomiting or constipation
7. Excessive barking

Destructive activity is the most common type of excitable behaviour that animal behaviourists are asked to treat. Many of these behaviours are self-rewarding. Chewing reduces tension. Physical activity does the same. Even urinating and defecating is self-rewarding because of the physiological effects these procedures have on the body. Because the actions are self-rewarding, separation anxiety can become self-perpetuating. In the dog's mind there is satisfaction from the activity. One fact concerning separation anxiety is often misinterpreted and that is that many owners feel that their dogs are destructive in their absence, or urinate and defecate out of spite, to get even for being left alone. This is not true. The dog's mind doesn't work that way. Quite simply, dogs are not capable of spitefulness. That is a characteristic behaviour of the primates, monkeys and us. The closest dogs come to spiteful behaviour is through jealousy and that is not applicable in the situation of separation anxiety. Dogs do not behave in these ways to get even with their owners. They don't act in these ways because of a lack of obedience or because they are dominant or even out of boredom. They do so out of anxiety.

Treatment of separation anxiety

Because so many separation anxiety problems are learned, there is an excellent prognosis for treating the condition. Simply by reducing the intensity of the relationship between you and your pet, by ignoring your dog for the half hour before you leave and again for a quarter hour after you return, there is a 40 per cent likelihood that the problem will correct itself without any further treatment. The other 60 per cent of cases need more complicated handling and there are a few points to first remember.

Separation anxiety is caused by separation from you, the leader of the pack. It isn't the same as boredom. This means that getting a pet for your pet won't work. Pets for your pets only works if dogs are genuinely bored. (Dogs that suffer from separation anxiety are vocal or destructive or hyperactive every time their owner leaves. Dogs that suffer from boredom are destructive or try to escape when left for prolonged lengths of time.)

Punishing a dog that suffers from separation anxiety can be counter-productive. In its own perverse way, it can be rewarding to the dog because he gets your attention. Similarly, tying him to the article that he has been chewing or sprinkling pepper or Tabasco sauce on the chewed objects can be counter-productive because it doesn't treat the root of the problem. All that this does is redirect his destruction else-where. If your dog has developed separation anxiety behaviour, follow these guidelines:

1. Reduce your dog's dependency on you by cooling your relationship with him. Become more remote. Stroke and pet him less and, if possible, have someone else feed him.
2. Try to avoid any destructive behaviour during the retraining period. This can be difficult. It might mean actually hiring a dog sitter for a short while or confining your dog to a cage or crate for short periods although this is best avoided if possible.
3. Retrain standard obedience behaviour – sit – stay – down (see Appendix).
4. Exercise your dog for a minimum of fifteen minutes at least twice each day.
5. Give your dog an acceptable item to chew, something that is entirely different to anything else in the house such as a rubber ring, Nylabone, tennis ball or rawhide chew. Don't give an old shoe. He will generalize permission to chew this to all other shoes that smell the same. Rub the item on your hands before you leave so that it retains your scent and only give it to him when you depart.

6. At the beginning of training, try to avoid separation anxiety behaviour whenever possible by taking your dog with you but, at the same time, start a routine of planned departures, perhaps even leaving a tape recorder running at home so that you can learn when his excitement level peaks and then disappears. In most instances, this is within twenty minutes of departure.
7. Calm your dog before departure for at least fifteen minutes by commanding 'sit-stay' and 'down-stay' and otherwise ignoring him.
8. All the moves in your mock departure should be as similar as possible to a real departure, always rewarding quiet behaviour but never rewarding anxiety. Then leave for a very short time and return, once more rewarding quiet and calm behaviour.
9. Only punish unwanted behaviour if you can hear it from outside and can burst in and startle your dog. Otherwise it is pointless to do so.
10. Sometimes it is useful to use a new stimulus during mock departures, something like a TV or radio left on. The dog associates this neutral stimulus with the happier circumstances of the new arrangement.

One specific manifestation of separation anxiety that is extremely common is excessive barking. Dogs that howl the house down are not necessarily doing it because of separation anxiety however. With some dogs, terriers in particular, there is a strong inherited behaviour to bark. Others will bark to defend their territory (warn the rest of the pack that strangers are near) or will bark when playing or chasing. Some barking is classic operant learning behaviour. The postman arrives, the dog barks, the postman departs. To the dog's mind it was his barking that made the postman leave. In a similar fashion dogs easily and readily learn that barking is a simple way of getting their owner's attention. My older dog barks once when she wants to come in from the garden. And, of course, many people actually train their dogs to 'speak', for food rewards.

Aside from these reasons however, excessive barking can be a manifestation of separation anxiety, a stereotyped behaviour in which the dog barks chronically when he is left alone. Retraining to overcome this problem is the same as for other separation anxiety behaviours and in fact applies to all causes of learned excessive barking. Don't reward barking even inadvertently. Shouting at your dog can actually be a reward simply because he's got your attention. Only punish the behaviour if your timing is perfect. You must punish while your dog is actually barking, not a few seconds after. Retrain to 'come', 'sit' and 'stay' and be consistent with training.

An unpleasant sounding device, the ultrasonic dog collar, can, in certain circumstances, be useful and necessary in overcoming excessive barking in dogs. This is a form of aversion therapy in which the dog associates the behaviour you wish to overcome with something that is unpleasant. The collar contains a vibration detector and when it is effectively used, the dog receives a high frequency buzz when he barks. This is unpleasant for him and when it is used sensibly, the collar can be very effective in overcoming the problem. There are other types of shock collars on the market, ones that are activated by the dog's owner and that give a short, low voltage shock. These should never be used to control barking because it's impossible to be consistently effective with timing their use. In fact, these collars should only ever be used under the guidance of experienced trainers and only for felonious canine crimes such as sheep chasing.

EXCITABILITY

Some dogs are constantly restless and excitable, not only when they are left alone. They react with uncontrollable excitement when the doorbell rings, they pull on their leads when exercised, dig relentlessly in the garden and never seem to tire, no matter how much exercise they receive. Owners will often call them 'high strung' or 'hyperactive' or even 'neurotic'. In some instances, this behaviour is inherited. In others, it is maternally imprinted and in both of these situations there is little that can be achieved in modifying the dog's behaviour.

We are responsible for much of our dog's activity and can sometimes create problems of excitability through our own behaviour. Some dogs get excited in the car because they have learned that the car journey ends with exercise and activity. Others get overexcited when visitors call because dogs are social animals and the appearance of any new person is an exciting event. We often inadvertently reward excitable behaviour by paying attention to the dog or even through mild discipline. And it is speculated that the physiological state of excitement is a reward in itself as it is in some humans, for example when they gamble.

Excitement in the car

Cars must be curious things to dogs. They can be frustrating because dogs can see interesting things from them, other dogs and green space, but can't get to them. They are also instruments of classical conditioning. Dogs readily learn to associate the car journey with the interesting sights along the way and the activity at the end of the journey. They are also instruments of learned behaviour. As the dog barks or gets excited, the car continues on its way, rewarding the dog with more interesting sights and sounds.

The owner's attempts to quieten the dog can also be rewarding.

Treating excitement in the car

1. As with almost all other forms of overcoming behavioural problems in dogs, the first thing to do is to retrain him to basic commands, 'sit', 'stay', 'come' and 'down'. Do this first in a neutral environment, such as the garden, and then later move to the car and repeat this training there.
2. Desensitize your dog to the car, initially by rewarding him for getting in and out of the car without either barking or being excitable and progressively by remaining calm when you get in, when you get in and start the engine and finally when you get in, start the engine and drive a short distance. During this training your dog should NOT be taken on any car journeys.
3. Reward calm behaviour and ignore excitable behaviour. For example, stop the car if he starts to become excited. Don't reward him with activity or attention. Wait until he calms down, then command 'down-stay' and proceed.
4. If he is small enough it might be helpful to put him in a cat basket so that he can't see out.

Excitement with visitors

The dog's mind enters a state of conflict when visitors arrive at home. He doesn't know whether they are members of his pack or not, friend or foe. If they are friends, then it's simply exciting that they have arrived. If, in his mind they are foes, then it's still exciting simply because their arrival has broken the monotony of the day but they are also a threat. This leads to ambivalent behaviour, curiosity and even pleasure that these strangers have arrived but also hostility to them, a hostility that can even lead to a dog biting while greeting. Hostility should be treated as you would for dominance behaviour, but if your dog simply shows excitement do the following.

Treating excitement with visitors

1. Reteach your dog to 'come-sit-stay-down' (see Appendix).
2. If his excitement starts with the doorbell, retrain obedient 'sit-stays' to

that sound first before proceeding to the visitor stage.

3. Disconnect your bell for a few weeks if your dog shoots from a cannon each time he hears it.

4. Ask your visitors to ignore your dog when they arrive. This can be hard to do. Some dogs just refuse to be ignored, supplicating visitors with mournful eyes, heads on laps, proffered gifts, and renditions of 'How much is that doggy in the window'. Equally, many visitors find it almost impossible to keep their eyes or their hands off the dog.

5. Keep a trailing lead on your dog so that if visitors call unexpectedly while you are in the middle of retraining, you can immediately and effectively control him by removing him to another room without touching him. (Touching can be potent reward.)

Excitement can show itself through exuberance, barking, jumping up and pawing for attention but equally, as we have discussed, it can manifest itself through fears and phobias.

SIGNS OF FEAR

Fight or flight

As has been mentioned, dogs will show their fear through fight or flight behaviour. They might growl or bite but equally might try to escape by running or hiding. Some will dig, chew or scratch to escape.

Freeze

Other dogs simply freeze when they are scared. They become stiff with fear. This can be a learned behaviour, something called learned hopelessness, and is more common than is frequently realized.

Excitement

Fear produces physiological changes in the dog. The heart and breathing rates both increase. Blood pressure increases. The pupils in the eyes dilate. Some dogs tremble. Others whine or bark. In extremes, some might urinate, defecate or empty their anal glands. All have poorer performance and poorer learning ability if they are fearfully excitable.

Activity

Dogs have a heightened startle response when they are fearful. They might be restless, pacing or running back and forth. Some indulge in stereotyped behaviour such as chewing their legs and others simply act hysterically. Many will actively seek out human company.

Treating fear

There are some general principles involved in treating fear, whether it is to animate or inanimate objects, to people or to thunderstorms.

1. Identify exactly what is causing fear and under exactly what circumstances fear is provoked. If a fear response is going to be treated, the stimulus of the fear must be one that can be controlled so that its intensity can be lowered and raised during training.
2. Desensitize and counter-condition your dog to the fearful stimulus by presenting it at a lower intensity, at an intensity that doesn't provoke a fear response, and reward his non-fearful behaviour with food and attention rewards. Food is a terrific antagonist to anxiety and fear.
3. Chronically exposing your dog to what makes him fearful, flooding his mind with the stimulus can work as long as it is a mildly fearful stimulus.
4. Once your dog has been trained to no longer be fearful of the stimulus, you should continue to periodically expose him to it so that the training is reinforced.

Treating fear of loud noises (thunderstorms and firecrackers)

Loud noises have both higher and lower frequencies than we can hear. Dogs are more sensitive to these noises, especially the low frequency noises that precede thunderstorms. If your dog has a fear of loud noises, make a recording of the noise and see whether the recording causes fear. If it does, you have a controllable stimulus to use in retraining. If the noise alone does not provoke fear, and you're really dedicated to overcoming the problem, add strobe lights, a sprinkler against the window and darkness if necessary to see if these factors affect his fear.

Using food rewards, train your dog to lie quietly on a rug and once he is fully trained, expose him to the frightening sound but at a very low level, perhaps even below your own hearing frequency. Reward him for showing no signs of fear. Gradually increase the intensity of the sound, boosting it every five minutes over a period of 30 to 40 minutes. Professor Don McKeown at the Ontario Veterinary College says that daily treatment periods of 40-50 minutes are best and that you should expect training to consume around 30 hours in total before a dog is desensitized to 90 decibel noise. Once your dog no longer shows fear to thunder, you should play your recording during the seasons when there are no naturally occurring thunderstorms, so that you reinforce his new behaviour.

Treating fear of people

Treatment of fear of people has already been outlined in Chapter Nine under fear induced aggression. Remember that it's very important to identify the exact circumstances of fear. It might only occur under highly specific situations.

STRESS INDUCED BEHAVIOURS

Barking is a classic stress induced behaviour in dogs but there are many other unusual manifestations of stress in which a dog behaves in a stereotyped but unusual way. These involve obsessive grooming, eating or body motions in which the dog carries out a certain behaviour out of context, fly chasing for example when there are no flies to chase. No one can really say exactly what goes on in the dog's mind when he carries out one of these behaviours, but they are some of the more unusual manifestations of stress that a dog can display.

In clinical practice, I have seen dogs that groom themselves so obsessively that they will lick down to the bone if not treated, others that whirl and circle, one that performed perfect figure eights. These behaviours can be caused by a certain type of epilepsy but in these circumstances, they were wholly anxiety or stress related activities.

Some Dobermanns are known to suck their flanks when agitated. Other dogs might appear to star gaze or twitch their skin. Some will bark rhythmically when anxious. All of these stress induced behaviours begin early in life. They can be avoided by controlling the circumstances that cause them, but in many cases the only effective treatment is with drugs (see Chapter Thirteen).

Because of their social nature, dogs are deeply influenced by their relationship with us and this is especially true with their fears and anxieties. Valerie O'Farrell at the University of Edinburgh has studied the influence of dog owner personality on dog behaviour and has produced some fascinating observations. O'Farrell says that it is only natural that dogs, with their exquisitely subtle and complex method of social communication, should be sensitive to their owner's actions, gestures and tone of voice. She carried out a survey of fifty dog owners, asking them questions about their dog's behaviour and asking them to rate their attitudes towards their dogs on various scales such as how much they missed them when they were parted or how upset they would be if anything happened to their pets. She also asked the owners about aspects of their behaviour towards their dogs, where they were allowed to sleep, how they were disciplined, what they ate and finally, she had the owners fill in a short personality inventory, specifically the Neuroticism scale of the Eysenck Personality Inventory, a form that has been shown to reliably differentiate between patients receiving treatment for neurotic illness and normal subjects. She, of course, also recorded general factual information on the owners of the name, rank and serial number variety.

O'Farrell's results are interesting because they confirm that the relationship owners have with their dogs affects the dog's behaviour but not necessarily in the way that is usually suggested. For example, dogs, whose owners are more anxious than normal, are more likely to show anxiety or stress related displacement activities such as excessive barking or chewing. Dogs, whose owners are emotionally attached to them, are more likely to show dominance aggression. O'Farrell says that the neurotic owner, because of his own greater anxiety, is more likely to induce excitement and conflict in his dog by the way he behaves towards it. She concludes that neurotic owners are more likely to punish their dogs inconsistently in ways that are likely to increase the dog's state of conflict. O'Farrell isn't the only psychologist who feels that we can affect the minds of our dogs in this way. At an international meeting on interrelations between people and animals, held in Monaco in November 1989, two psychiatrists and a veterinarian from Japan presented a research paper that came to the same conclusion. In the case of the Japanese study, the researchers felt that there was a relationship between the behaviour of the owners and psychosomatic behaviour in their dogs. Many owners look upon their dogs as extensions of themselves. They project feelings that they personally find difficult to express in their own personalities onto their dogs. This can produce conflict in the dog's mind and consequent anxious behaviour. Once more, it is we who are modifying the dog's mind and we are unwittingly doing so because of our own vanities and insecurities.

Chapter Eleven

Social Behaviour – Pack, Sex and Maternal Activity

PACK BEHAVIOUR

Social behaviour in wolves is based, as has already been discussed, on a dominance – subdominance – submission basis. A hierarchy develops and is maintained within the pack through ritual behaviours. The dominant wolf shows confident postures, makes confident sounds and marks his territory with confidence. He walks with stiff legs and high, moderately wagging tail. He sniffs nose to nose, then nose to genitals. The subdominant tag along in a contented fashion with more subdued behaviours while the submissive perform ritual passive behaviours, grovelling with the head and tail down, crawling on the belly and trying to lick the lips of the dominant wolf. The supersubmissive might urinate in abject submission at the same time.

Dominance is maintained through ritual behaviours.

Pack behaviour in dogs is less complex but, because rules can't always be followed, the behaviour is less predictable. Few dogs ever gain the ability to form a pack. Only dogs that have reverted to the wild do so, dingos and pariah dogs. Free ranging dogs on the other hand, domesticated housedogs that are allowed to roam free, are a common problem in any large city.

These free ranging dogs, as I have previously mentioned, have been studied in considerable detail and the most obvious common finding in all studies is that, as concentrated as the populations are, the dogs never form social packs with dominance hierarchies. Most of these dogs are lonesome travellers although two neighbouring dogs will often range together. The only time that dogs congregate is when there is a female in estrus and even then it's rare for there to be more than seven dogs following one estrus female.

When Thomas Daniels conducted his observations of free ranging domestic dogs in Newark, New Jersey, he observed that strays were actually quite rare. Most of the dogs were owned and there were three free ranging males for each female. Dogs that were familiar with each other rarely fought and in fact in most facets of their behaviour, it would be difficult to describe them as a pack. The word 'pack' implies a degree of mutual cohesion but the dogs in Daniels's study simply weren't very sociable, probably because they didn't have the need to defend a food source or defend themselves against predators.

Daniels observed no territorial behaviour in his Newark dogs, nor did Alan Beck in his Philadelphia dogs, nor for that matter did Ian Dunbar in his more up-market Berkeley, California dogs. What all of these researchers observed was that free ranging dogs aren't promiscuous.

SEXUAL BEHAVIOUR

As we already know, the male pup's brain is 'masculinized' near birth by a surge of the male hormone testosterone. The mind of the female pup remains 'neutral' and this is what we consider to be normal for bitches. At puberty, the female experiences a surge of the female hormone estrogen. This sudden production of estrogen has a dramatic effect on her responses and at the appropriate moment, she will allow mating. Afterwards, whether she has been mated or not, she will experience a constant and prolonged episode of production of the hormone of pregnancy, progesterone. This too will dramatically affect her behaviour, as I have already described in Chapter Four. Sexual behaviour involves hormonal influences, spinal reflexes, active use of the brain and previous experience and it does not reveal itself only through overt sexual behaviours such as urine marking or sexual aggression. There are other subtler behaviours that are influenced by the sex of the dog. In an extensive survey carried out from the University of California, authorities on animal behaviour performed a gender analysis of thirteen common canine behaviours and came to the conclusion that females were easier to obedience train, were easier to housebreak and were more demanding of affection.

They found no sex difference in watchdog barking, excessive barking or excitability, but found that male dogs were more playful, more destructive, more likely to snap at children and defend their territory, more active, more aggressive towards other dogs and more dominant over their owners. They were graded in the following percentile rank:

Obedience training	TRAITS MORE LIKELY	55	
Housebreaking ease	TO BE SHOWN	45	
Affection demand	BY FEMALES	25	
Watchdog barking		0	
Excessive barking		0	
Excitability		0	
Playfulness		15	TRAITS MORE
Destructiveness		20	LIKELY TO BE
Snapping at children		20	SHOWN BY
Territory defence		35	MALES
General activity		40	
Aggression with dogs		65	
Dominance over owner		70	

The sex of the dog has a significant influence on many aspects of the dog's mind but it is, of course, most influential on sexual behaviour.

Females give generous clues to dogs that they are approaching ovulation. They can act playfully with dogs, making body contact and running with the male but avoiding actual mating. Play wrestling is common with the female extending her forepaws while keeping her haunches up in the air and her head to the side. Michael Fox calls this 'play-soliciting'. They might urinate more frequently, leaving odour clues to their approaching ovulation. Some bitches will try to mount other dogs and others will simply bark more than they normally do. They can show increased activity and increased aggression as they approach ovulation, finally standing for the male and allowing him to mount by placing her tail to the side and displaying her vulva.

Male dogs, in turn, pursue and investigate females in season. They sniff and lick the female's anogenital region and then urine mark nearby. They too play wrestle in the same 'play-soliciting' fashion, but almost always with the tail and ears erect. Male dogs stand squarely in front or to the side of the female in an alert fashion. This leads to their pushing against the female with the neck, shoulder or hip. Following this, the dog places his neck or even his forefoot on to the shoulder or the back of the female and if there is no resistance, he might lick or nibble that region. This is followed by mounting, clasping with the forelimbs, pelvic thrusts and copulation, followed by the dogs 'tying' together for any time from five to fifty minutes. It's thought that the tie evolved to prevent other dogs from immediately mating with the female.

Females are selectively receptive to male dogs, readily admitting some males and rejecting others. This receptivity is based on familiarity and in a semi wild or free ranging environment, unfamiliar dogs are always unsuccessful at mating. Familiarity is more important than the dominance of the male, although the presence of estrous females is the only circumstance that always bring free ranging dogs together in such a way that they form a hierarchy or engage in fights. In street observations, persistence at attempting to mate didn't increase the likelihood of an unfamiliar dog mating with the female in season. Rank and size didn't matter either. The females were receptive to specific males they were familiar with rather than to the most dominant dogs.

Of course, there are dramatic individual differences in sexual behaviour both in free range dogs and, more practically speaking, in house dogs, and these can be genetic rather than learned.

SEXUAL BEHAVIOUR PROBLEMS

Excessive mounting – masturbation

Male pups will naturally mount other pups or other objects such as cushions or even arms, but most will outgrow this behaviour. In pups, it is usually simply a sign of dominant behaviour. Adult male and female dogs might continue mounting behaviour and in many instances, they do so when they are excited or stressed. In these circumstances, mounting can become a stereotyped repetitive behaviour that is carried out to diffuse agitation or excitement.

In other circumstances, masturbation with cushions or with people occurs at puberty in dogs that have no social contact with other dogs. It is quite simply a substitute activity and can occur in neutered or entire males and females. Finally, mounting and masturbation can be an attention getting mechanism, conditioned by the owner. In all of these instances, modification of the behaviour follows the same principles. Re-teach obedience commands. Remove the objects that the dog is using. Increase his mental and physical activity to give his natural energy another outlet. Discipline him when he mounts or even eyes your visitor's nubile knees.

Dogs can be quite sophisticated in their attention getting behaviours. It's not that they have devious minds but rather that they are exquisitely sensitive to nuance and detail. If a dog is mounting visitors' legs or trails about after you in the house leering and flashing, any type of retraining should avoid physical contact with him as this in itself is probably a potent reward. In these circumstances, it's best to leave his lead trailing from his collar and to isolate him in a quiet room for about three minutes each time he tries to mount you, the kids or your visitors. I once described this method of retraining to the owners of a salacious Dachshund who, on the first day of its implementation, had to isolate the dog 84 times. But three weeks later, his mounting behaviour had reduced to nil.

Lack of sexual interest

Lack of interest in sex can be of a temporary or permanent nature. The wild ancestor of the dog, the wolf, is basically a monogamous animal, pair bonding for life. Selective breeding for a thousand generations has reduced monogamy in dogs to the state it is today but there can still be a strong preference to certain mates. Familiarity can also diminish interest in mating. A male and female raised and kept permanently together might be less likely to mate than the same dogs separated for a week or ten days before estrus.

Lack of interest in sex can be permanent because of the early imprinting of the dog onto humans. In these circumstances, where the dog has never been properly socialized to other dogs, mating is unsuccessful simply because of anxiety in the presence of a member of their own species.

In the context of sexual behaviour I would like to mention once more the advantages and disadvantages of neutering dogs, if only because it is the most common operation carried out in veterinary medicine and one over which there are the most misconceptions.

Castration

Castrating a male dog is likely to have a significant effect on his mind. You should expect to see:

Less aggression towards other male dogs.

Less inclination to try to dominate the owner or other members of the household.

Less urine marking inside the house or in other houses.

Less inclination to mount other dogs, visitors' elbows and knees or teddy bears.

Less inclination to roam away from home.

There is a 50:50 chance that castration will alter these canine behaviours if a dog has already developed them. In half of the cases, the improvement will be immediate and in the other half, it will take longer. Because the male dog's brain is masculinized from birth, because of that early surge of influencing male hormone, it doesn't matter whether a dog is castrated before or after puberty. In other words, there is no need to castrate a dog before puberty, to castrate him preventatively in order to try to avoid unpleasant sex hormone related behavioural problems. If a troublesome problem does develop, it is just as likely to respond to castration after puberty and after it has developed as it would to preventative castration.

Castration will NOT make a dog calmer or less destructive or better with children. It will not decrease excessive activity either. If there are behavioural indications for castration, it is as likely to be clinically effective in one dog or another regardless of his age or the degree of his objectionable behaviour. Animal rights activists in some countries feel that castration is an unfair intervention in animal behaviour and in Sweden, it cannot be performed on a dog without there being a good clinical reason. While applauding anyone's concern for animal welfare, I feel that this is a hopelessly anthropomorphic attitude, totally out of harmony with our need to control the size and activity of the animal population for which we have a responsibility. Curiously, because of the wording of the legislation in Sweden, spaying dogs in that country falls under the same restrictive covenant.

Spaying

Spaying a female dog does not result in any significant degree of behavioural change and the reason is quite simply that because females are hormonally active only twice a year, their minds are not constantly under hormonal influence. Unlike male dogs, their brains are NOT femininized at birth. The first sex hormone to surge through their bodies does so when they reach puberty.

Females are not normally spayed for behavioural reasons but rather to prevent them from conceiving. The operation is carried out to eliminate the twice yearly vaginal discharge and the possibility of an unwanted pregnancy. Although there is no scientific evidence to indicate that the female sex hormone has a significant long term effect on the dog's mind, there is ample anecdotal evidence that this can sometimes be the case. I'm a source of some of that anecdotal evidence. In clinical practice, I see the females undergo behavioural changes when they experience the hormonal changes associated with estrus and pregnancy or even phantom pregnancy. These changes can include an increased snappiness or reduced general activity and they can become perpetuated after the hormonal influence ceases to stimulate them. This is why I feel that if you know for certain that you do not plan for your female dog to have pups, then she should be spayed when she is physically mature and before her first season. There are occasionally physical reasons why this should not be done but these are uncommon.

MATERNAL BEHAVIOUR

Dramatic hormonal changes occur as whelping approaches and, as I have discussed in Chapter Four, these changes have a considerable effect on the pregnant mother's mind. As parturition nears, both her estrogen and prolactin hormone levels increase. For birth to take place, the mother's estrogen level has to supersede her progesterone level. From a behavioural point of view, she will spend an increased amount of time licking her nipples and vulva. She will be increasingly restless and her appetite will decrease. At this time some females can become increasingly wary of strangers and can become quite aggressive if disturbed.

Her behaviour at birth can vary considerably. Her breathing can vary from panting to slow deep respirations. She might even close her eyes. Her contractions can vary in intensity too and there is usually a period of rest between deliveries although two pups are often delivered close together followed by a longer rest period of anywhere from ten minutes to two hours. Throughout her delivery, the female usually doesn't move around very much. If she does get excited, as many small dogs such as Yorkshire terriers do, she can actually inhibit her contractions. This is once more a hormonal condition. Excitement produces adrenalin and this inhibits contractions of the uterus.

All these changes, an enlarged abdomen, mammary gland development, milk production, nest building, increased nervousness and aggression, mothering toys, attempts to nurse and even labour, can occur in the false pregnancy that always follows estrus in females. In mind and body, the female thinks she is pregnant and the entire episode only comes to an end when her progesterone level, the hormone of pregnancy, falls back down to its resting level.

It can be lucidly argued that we keep our pet dogs in unnatural circumstances, in which their natural behaviours are diverted into dead end activities such as masturbating with cushions, mothering green felt frogs and forming packs with essentially disinterested members of another species, us. This is all true but the practical fact is that this is now the natural circumstances of dogs. The companion dog is a man-made animal, no longer the creature of the wild following millenia of evolved behaviour patterns. We have modified the dog in countless ways and in doing so, we have created genetic lines of dogs with different minds and even cultures. These are the breed differences and the subject that we should look at next.

Chapter Twelve

Breed Differences in Behaviour

Bulldogs were bred to attack and hang on the noses of bulls. Anatolian sheepdogs were selectively bred to guard sheep in the mountains of Turkey. The Chihuahua was bred to act as a hot water bottle. The St Hubert hound was bred to follow scent better than any other dog. Mastiffs were bred as dogs of war. Borzois were bred to outrun wolves. Irish water spaniels were bred to retrieve game from the most inaccessible of watery locations. Throughout the history of our relationship with them, we have selectively bred dogs to perform certain functions. Unwittingly at first and then intentionally afterwards, for a thousand generations, we have been tampering with the dog's mind.

We still do so today. Throughout the world, there are at least 400 different breeds of dogs and as I have described in Chapter One, there are between 30,000 and 50,000 genes which govern and monitor the individual features and characteristics of each one of them. In nature, only those animals best adapted to survive on their own will reproduce but this is not the case with domesticated animals such as the dog. Because of our intervention, mutations that would normally disappear, mutations such as heightened aggression or gentleness of mouth, have been allowed to continue. Creating the best possible environment for our dogs and feeding the best foods have also allowed negative mutations to continue. Breeds with pushed in faces such as the Pekingese would disappear within a generation if left to natural selection.

Dogs have been bred for a range of activities including:

Hunting
Attacking other dogs or animals for sport
Attacking people
Farmyard work
Guarding and watchdog activity
Herding
Racing

Sledge and cart pulling
Scene following
Acting as comforters

All of these behaviours and abilities are already present in the genetic make-up of dogs. Originally, dogs were bred for performance characteristics, not morphological characteristics. Dogs were bred to behave in certain ways. Inbreeding perpetuated these behavioural traits, these states of mind. The various shapes, colours and sizes of dogs developed on a secondary basis. The poodle is a classic example. The 'pudel' was originally bred as a German working dog with a water resistent coat for working in marsh and pond. It was not bred for the fifteen different colours and sizes that are bred for today, but rather for its utilitarian function. Konrad Lorenz in 'King Solomon's Mine' says that 'selective breeding aimed at physical features is not consistent with breeding aimed at emotional features'. In this instance he is only partly correct for when we inbreed for physical characteristics, we do breed for emotional features only they are not neccssarily the emotional features we actually want. Breeding for physical characteristics does mean that we are also breeding for emotional homozygosity but that can be quite different from breeding for performance.

Some breeds have been selectively developed for enhanced scent following ability.

The consequence is that hounds such as Beagles and Bloodhounds, bred for brilliance at following scent, have baggy lips perhaps to help take in more taste and smell. But they are also surprisingly difficult to actually train to do anything. They simply feel impelled to follow scent. Bird dogs such as Hungarian Viszlas and Golden retrievers have been bred for soft mouths and inhibited aggression. To act in such a way they must be highly trainable, Similarly, pointers have been bred to hunt but not to carry out the final kill, once more inhibited aggression. The same rule applies to herding dogs such as Collies. They stalk and herd but their aggression is inhibited

Shepherd breeds such as the Maremma have been developed to look like the sheep they protect.

Guarding herd dogs had two roles, to guard and protect. The German shepherd is the classic example which is why it is so popular and successful as a dual purpose dog today. Other breeds were bred solely as guards, for heightened independent aggression. Great Danes, Mastiffs and Irish Wolfhounds all originally fell into this category and none are as trainable as German shepherds. Fortunately they are now bred for docility but in both Danes and Mastiffs, throwbacks do easily occur.

Scott and Fuller at Bar Harbour, Maine carried out the first considerable scientific observations on breed differences in behaviour. They studied emotional reactivity, trainability and problem solving ability in Basenjis, Beagles, Cocker spaniels, Shetland sheepdogs and Fox terriers. Emotional reactivity was measured by the amount of distress barking carried out by the dog, tail wagging, heart rate and respiratory rate.

In their experiments, the experimenter entered the room housing the dog, spoke softly and left. On other occasions he would enter, grab the dog by his muzzle and give a shake or he would ring a bell or he would administer a shock to the dog's leg. In emotional reactivity they concluded that the Fox terriers, Basenjis and Beagles were more reactive than the Shelties and Cocker spaniels.

Trainability was examined through lead training and the dog's willingness to jump down on command. In these tests Cocker spaniels were best, Basenjis and Beagles worst and the Shelties and Fox terriers middling. Finally they looked at problem solving. This involved detour tests, maze tests and manipulation tests and here the hunting breeds all did far better than did the Shelties. Scott and Fuller concluded that all breeds appear to be quite similar in pure intelligence but that there are enormous differences both within and between breeds for developing certain capacities.

Aggressive behaviour, for example, has been dramaticaly modified in different breeds and can be highly selected for. We enhance or diminish it but can't alter the fact that all breeds fight in the same way. It is simply easier to stimulate this behaviour in some breeds. But how much of this aggression is inherently in the mind of the dog and how much of it is created through our intervention?

The most troublesome breed of dog today is the American Pit Bull terrier. In North America, more human deaths and serious injuries are caused by this breed than by any other. They have been selectively bred to attack and to hold on tenaciously. They have jaw muscles that can be inches thick with a crushing power of hundreds of pounds per square inch. All information points to the conclusion that these dogs are simply killing machines. But as with all other aspects of the dog's mind, the situation is not necessarily as it outwardly appears.

When twenty human fatalities from Pit Bull terrier attacks were studied, several common denominators were observed. Ten of the twenty lethal dogs were owned by males between 20 and 25 years old. Eleven of the owners of these twenty dogs had criminal records and seven of these had criminal records for violence. Eleven of the dogs showed evidence of physical abuse.

There is no question that the Pit Bull terrier has been selectively bred for heightened and tenacious aggression but the problem with the breed is not simply one of genetics. It is in fact a human problem in that, for whatever the reasons may be, these dogs are bred and trained, often trained with cruelty and abuse, to attack and to not let go. Many towns in the United States and Canada have passed local ordinances restricting the keeping of Pit Bull terriers. One of the first towns to do so was Lynn, Massachusetts. A short time after the ordinance was passed, a local veterinarian reported that whereas he had 300 Pit Bull terriers in his records before the law was passed, he now had 200 and no pups.

However, the number of Rottweilers in his records jumped from 20 to 150 in the same period.

The mind of the dog is highly flexible and is open to drastic modification through learning and experience. Pit Bull terriers, together with German shepherds, Rottweilers, Dobermanns, Akitas and Staffordshire Bull terriers all have a heightened genetic potential for aggression but this is exaggerated through early training. Border collies are born with a predisposition to herd sheep but they also need training and experience to meet the full potential of their abilities. Similarly, a fear of noise in Collies, flank sucking in Dobermanns and attachment related problems in Labradors all have an inheritable breed component to them, but are accentuated through early learning. Even behavioural tendencies with an apparently strong genetic basis can be altered through experience and early training. In the early 1960s, researchers in the United States observed that aggressive strains of Wire-haired Fox terriers could not be reared together in normal litters because subordinate pups tended to be killed or mutilated by their dominant siblings. These latent aggressive tendencies never developed, however, when litter mates were isolated at an early age and hand reared until weaning.

Although most of the breeds we have today were originally bred for specific purposes, their roles can now be considerably different. The Hungarian sheep guarding dog, the Komondor, is being used with varying success in the United States to fulfil its original guarding function, only now it guards against coyotes. The Belgian Malinois on the other hand is used on the Mexican border by the American Drug Enforcement Agency to intercept illicit drugs. In their first year alone, two Malinois, obedience trained in Dutch incidently, intercepted $135,000,000 worth of narcotics. Their handlers keep red rubber balls in waist pouches and reward their dogs with the ball when they make a drug find. Apparently, frisbees for the dogs come on the government payroll too.

Through controlled breeding of dogs we have accentuated certain canine traits and through inbreeding we have created breeds with specific behavioural repertoires. We have created breeds to act consistently, for similarities of mind, and to follow certain behaviour patterns. Today we perpetuate breeds of dog with consistent morphological characteristics and Ben Hart has studied this in fascinating detail.

Ben Hart is a veterinarian and professor of animal behaviour at the School of Veterinary Medicine at the University of California. His wife, Lynette Hart, is a sociologist and director of the Human-Animal Program at the same university. In 1985, in the Journal of the American Veterinary Medical Association, they published the results of the most extensive survey ever carried out on the minds of different breeds of dogs. The Harts selected thirteen different canine behaviour character- istics and then asked ninety-six experts, half of them practising veterinar- ians and the other half dog obedience judges, to rank different breeds according to these characteristics. Because there are well over 100 registered breeds in the United States, the Harts restricted this evaluation to the fifty-six most popular breeds and asked each authority to rank seven breeds.

The reliability of the characteristics to correctly differentiate between breeds was determined using a statistical method called the F ratio, the higher the F ratio the more reliable the characteristic. These are the behavioural characteristics the Harts used and their F ratios.

Excitability	9.6
General activity	9.5
Snapping at children	7.2
Excessive barking	6.9
Playfulness	6.7
Obedience training	6.6
Watchdog barking	5.1
Aggression towards other dogs	5.0
Dominance over owner	4.3
Territorial defence	4.1
Demand for affection	3.6
Destructiveness	2.6
Ease of housebreaking	1.8

As you can see, behaviours such as excitability and general activity are reliable indicators of differences in the minds of different breeds while destructiveness and housebreaking are less influenced by genetic con- siderations and are more a product of training and the environment.

To get their results, the investigators asked their experts specific questions. Excitability, for example, was explored with this question:

'A dog may normally be quite calm but can become very excitable when set off by such things as a ringing doorbell or an owner's movement toward the door. This characteristic may be very annoying to some people. Rank these seven breeds from least to most excitable.'

Using a ranking that divided all fifty-six breeds on a scale of one to ten where excitability increased with number rank, these were the Harts published results:

Bloodhound
Golden retriever
Newfoundland
Akita
Rottweiler
Chesapeake Bay retriever
} 1

English springer spaniel
Dalmatian
Cocker spaniel
German shepherd
Shih Tzu
Samoyed
} 6

Labrador retriever
Australian shepherd
Great Dane
Old English sheepdog
Alaskan Malamute
} 2

Scottish terrier
Weimaraner
Dachshund
Pug
Airedale terrier
} 7

Saint Bernard
Boxer
Dobermann
Viszla
Collie
Bulldog
} 3

Irish setter
Maltese terrier
Pomeranian
Lhasa Apso
Shetland sheepdog
Boston terrier
} 8

Chow
Brittany spaniel
Basset hound
Norwegian elkhound
Afghan hound
} 4

Chihuahua
Silky terrier
Pekingese
Toy poodle
Miniature poodle
} 9

German shorthaired pointer
Welsh Corgi
Standard poodle
Bichon Frise
Keeshond
Siberian husky
} 5

Yorkshire terrier
Cairn terrier
Miniature schnauzer
West Highland white terrier
Fox terrier
Beagle
} 10

This new classification of breeds of dogs based upon their temperament rather than their morphology or their original purpose was a radical departure from previous methods of classification and listing for each of the thirteen traits, listings similar to this one, were published by the Harts in 1988 in their book 'The Perfect Puppy'.

The consequence was that for the first time, it was possible to look at dog behaviour on a comparative basis and to see just how different breeds can really be. In their original scientific article, the Harts published behavioural profiles for the Basset hound and the Miniature schnauzer and they looked like this.

Trait	Basset hound	Miniature schnauzer
Excitability	1	10
General activity	1	10
Snapping at children	1	10
Excessive barking	4	10
Affection demand	1	7
Territorial defence	1	10
Watchdog barking	1	10
Aggression with dogs	4	10
Dominance over owners	3	10
Obedience training	1	7
Housebreaking ease	1	3
Destructiveness	2	8
Playfulness	1	10

This was as clear an example as any of the influence of genetics on the dog's mind. As you can see, these behavioural traits are clustered and this was their final consideration in the ranking of breed behaviour. Using a statistical method called factor analysis, the Harts divided these thirteen traits into four general predispositions of the dog's mind which they called:

Reactivity   (= excitability, general activity, snapping at children, excessive barking, demand for affection)

Aggression   (= territorial defence, watchdog barking, aggression with dogs, dominance over owner)

Trainability   (= obedience training, housebreaking ease)

Investigation   (= destructiveness, playfulness)

Then, using another statistical method and with the help of their ever ready computer, they performed a cluster analysis by which they assigned breeds of dogs, according to their behavioural profiles, to specific groups. The surprise was how closely these new behavioural groups corresponded to the conventional grouping of dogs into hound, sporting, working, terrier and miscellaneous groups.

Cluster analysis of dog behaviour produced this new classification of breeds according to their behaviour:

Cluster 1:
   High reactivity–low trainability–medium aggression
1. Lhasa Apso   7. Beagle
2. Pomeranian   8. Yorkshire terrier
3. Maltese terrier   9. Weimaraner
4. Cocker spaniel   10. Pug
5. Boston terrier   11. Irish setter
6. Pekingese

Cluster 2:
   Very low reactivity–very low aggression–low trainability
1. English bulldog   4. Bloodhound
2. Old English sheepdog   5. Basset hound
3. Norwegian elkhound

Cluster 3:
   Low reactivity–high aggression–low trainability
1. Samoyed   6. Boxer
2. Alaskan Malamute   7. Dalmatian
3. Siberian husky   8. Great Dane
4. Saint Bernard   9. Chow
5. Afghan hound

Cluster 4:
   Very high trainability–high reactivity–medium aggression
1. Shetland sheepdog   5. Bichon Frise
2. Shin Tzu   6. Standard poodle
3. Miniature poodle   7. English springer spaniel
4. Toy poodle   8. Welsh Corgi

Cluster 5:
   Low aggression–high trainability–low reactivity
1. Labrador retriever   6. Chesapeake Bay retriever
2. Hungarian viszla   7. Keeshond
3. Brittany spaniel   8. Collie
4. German shorthaired pointer   9. Golden retriever
5. Newfoundland   10. Australian shepherd

Cluster 6:
   Very high aggression–very high trainability–very low reactivity
1. German shepherd   3. Dobermann
2. Akita   4. Rottweiler

Cluster 7:
   Very high aggression–high reactivity–medium trainability

1. Cairn terrier   6. Dachshund
2. West Highland white terrier   7. Miniature schnauzer
3. Chihuahua   8. Silky terrier
4. Fox terrier   9. Airedale terrier
5. Scottish terrier

Twenty years of active veterinary practice during which I have observed dog behaviour and listened to the complaints or boasts of dog owners has resulted in my having developed firm opinions about the behaviour of the more popular breeds of dogs. It came as a pleasant surprise when I read the Harts' results to find that my anecdotal experience was in almost complete harmony with their findings. Of course, there are national or even regional differences. In Britain, for example, quarantine regulations mean that it can be difficult to develop a foreign breed from anything other than a few imported animals. The consequence can be that breed peculiarities might arise in this population that may not arise elsewhere. Forced inbreeding can mean that the dog's mind can form along different tangents in different parts of the world. An example is the West Highland white terrier. In the Harts' American survey, this breed came just behind the Miniature schnauzer in setting the record for excitability, general activity, snapping at children, destructiveness, aggression towards other dogs and excess barking. In diplomatic tones, the Harts say that this breed 'might very well challenge the ingenuity and patience of owners who aren't assertive around dogs'. In the UK where there is a larger genetic pool, there are thousands of Westies that fall into that category but in my experience, there are an almost equal number that have a much lower aggression and greater trainability, dogs that behave more like Shih Tzus and Poodles.

   The genetics of behaviour is an accepted fact. Rene Descartes' comment that animals are machines has a new connotation today as we come to realize that all animals, including us, are exquisitely developed machines that function on the basis of the genetic information imprinted into our circuitry. The Harts' survey of some of the behavioural characteristics that make up the dog's mind has been an excellent first step towards a more scientific examination of the genetics of canine behaviour.

Chapter Thirteen

The Mind of the Ill and the Elderly

The dog's mind is crafted by the information it receives. Any change in his ability to receive useful information and to process it will affect his behaviour. Both the ageing process and illnesses interfere with sensory perception, hormone production and the ability to communicate, all of which lead to sometimes dramatic changes in a dog's behaviour.

The Ageing Process

Just as La Rochefoucauld said of us, with advancing years dogs too become wiser and sillier. Memory decays in old dogs and they do everything more slowly. That includes thinking. We all know, however, that the onset of the ageing process and the speed of its progression varies quite dramatically from dog to dog. In clinical practice, I see eight year olds that quite simply are elderly in their behaviour and I also see fifteen year olds that still have a youthful inquisitiveness and responsiveness. There is almost undoubtedly a biological clock that governs the ageing process in dogs and this is genetic in its origins. This is why the life expectancy of some breeds is dramatically longer than others. A miniature poodle should be expected to live for fifteen to sixteen years yet a Cavalier King Charles spaniel, a dog of almost similar size, has a life expectancy of eleven or twelve years.

To understand how the dog's mind can change with age, it is useful to briefly look at how his information technology is modified through advancing years.

There is a widely held belief that a biological clock controls the ageing process and that clock, genetic in origin, does so by influencing the body's hormone systems. The clock itself is possibly located in the hypothalamic area of the brain, the area that is known to control growth hormone as well as the activities of many of the body's hormone producing glands. Naturally, the environment will have a dramatic

*The ageing process is probably under the control of a hormonal biological
clock.*

effect on the ageing process but all that the environmental influences
will do is either speed up or slow down a genetically predetermined life
span. In other words, ageing and the changes in behaviour that accompany it can be influenced by environmental factors but the biological
clock associated with death is genetic and therefore immutable. By
improving the environment we can prolong life but there is a maximum
life span for a dog that is written in his genes.

Many of the physical changes of ageing that occur in nerves and the
brain are already known. Old dogs have lighter brains than young dogs
(25 per cent lighter) although this does not necessarily mean that they
have fewer nerve cells. Although nerve cells are undoubtedly lost with
age, the often quoted fact that from the onset of maturity we humans
lose 100,000 cells a day is a gross exaggeration. What does happen is
that the structure of nerve cells in the brain breaks down. Filaments
from nerve cells contract and they lose some of their contacts with other
cells. It is as if part of the magnificent wiring system simply burns out.
This means that the chemical factory that is the brain no longer
functions at its previous level of efficiency. Chemicals aren't transmitted
from one cell to another or are not even packaged properly. It is as if
millions of mature and lush weeping willows have been pruned back to
their trunks.

Jacob Mosier, Professor of Medicine at the veterinary school at Kansas
State University, says that the signs of senility in dogs are similar to

those in man. Older dogs have poorer reflexes and suffer sensory losses both of which affect his behaviour. With the pruning of the cells of the brain that occurs with age, transmission time takes longer. In a healthy young dog, information is conducted through the nervous system at around 225 miles per hour. In the elderly dog, this can slow down to perhaps 50 miles per hour.

Within the brain itself, most of the nerve cells, by one estimate 95 per cent of them, are 'interneurons', cells that amplify and refine signals that are received by the brain. With age, these cells when stimulated remain stimulated for abnormally long periods. The consequence is that the first piece of information they receive temporarily puts them out of action. They can't amplify or refine any following information. The result is that a dog's short term memory is affected and this is why it is more difficult to teach an old dog new tricks. It's also why his response time is increased.

Other changes affect brain function too. With age, the blood vessels of the brain lose their elasticity and the lungs become less efficient. Remember that the brain takes almost 20 per cent of the blood that the heart pumps out. The inefficiency of the blood vessels and lungs means that the brain becomes chronically starved of oxygen and this has an effect on long term memory. At the same time, ageing results in a thickening of the membrane that surrounds the brain, the meninges. It can even become ossified, hard and brittle. Tiny haemorrhages can occur around blood vessels, little microhaemmorhages that destroy only a few nerve cells but that still affect the mind and behaviour of the dog. A consequence is that older dogs can become irritable when disturbed, can become slow to obey commands and can ultimately have problems with orientation and learning. I once had a golden retriever that I ultimately euthanized when she was close to seventeen years of age. She never showed any signs of irritability but otherwise her senile changes were those of reversion to a puppy like dependence. Although she had walked without a lead throughout her life, for her last year she was visibly more relaxed when she was on a lead. She preferred the presence of her 'family' and would seek us out, only to lie down and fall asleep where we were. Her previous daily activities became ritualized. She asked to go to the park in her usual way, not because she wanted exercise but only because it was a behaviour that was imprinted in her mind. Once at the park, she would leave the car, have a fifteen second roll then turn around, return to the vehicle and climb back in. If, in her last few months, I had carried out an EEG on her, I probably would have seen a typical puppy pattern of electrical waves.

Caleb Finch, in his book *Biology of Aging*, says that the brain controls or at least influences the ageing process through its control of the body's hormones. He has described how the chemical factory becomes less efficient, producing fewer neuroendocrines and specifically how its production of the neuroendocrine dopamine diminishes. There certainly seems to be a relationship between dopamine production and life span in mice. The ageing process and the changes of behaviour that accompany it involve a complex and still unknown relationship between the brain and the body. The relationship is certainly dynamic and is mediated through the hormone system with the hypothalamus interacting with the pituitary gland which in turn interacts with the various glands around the body. With age, the thyroid gland decreases in weight, the testicles might produce less testosterone and the ovaries start to shrink. At ten years of age, the adrenal gland no longer responds as quickly to chemical messages from the brain as it did at five years of age. One small change in one part of the system can result in a cascade of events that might dramatically alter the dog's behaviour.

At the same time that the physical functioning of the brain, hormone and nervous system change, so too does the sensory system that feeds information into the dog's mind. The dog's hearing changes. He is no longer as sensitive to high notes. There is connective tissue change to the cochlear vessels, degeneration to the ganglia and loss of hair cells. In some breeds, the Labrador and golden retrievers for example, arthritis of the ears can cause deafness.

The eyes change too. Old eyes lose both types of retinal cells, the rods and cones. The consequence is that the retina becomes disorganized. At the same time, the dog's lenses lose their elasticity and become foggy with fibre tissue, a condition called sclerosis. The result is that they can't focus as well as they used to and see through a haze.

Taste is affected too. It seems that both sweetness and saltiness lose their intensity as do flavours, probably as a result of incomplete cell replacement. Smell is the sensory ability that lasts longest. In one curious experiment that has direct theoretical application to the dog, it was noted that people who used their sense of smell extensively throughout their lives retained their sense of smell longer. In this experiment perfume testers in their seventies, people who had not been originally chosen for the job because of their scent faculties, were found to have less of a decline in either the threshold at which they could smell or in their ability to discriminate odours. When seen in the light of other research, this has an immediate application to the management of the mind of the elderly dog.

We know that a reduction of oxygen to the brain results in impaired brain function, that long term memory is impaired. But experiments have equally shown that increasing the supply of oxygen to the brain in old dogs will improve long term memory quite dramatically. The immediate conclusion is that healthy and active dogs will experience fewer or slower senile changes to the brain than will unhealthy and inactive dogs. Exercising your senses, as the perfumers did, delays the natural deterioration of ageing. A recent experiment using rats has helped to explain why.

William Greenough at the University of Illinois moved a number of elderly overweight rats that had lived their lives in featureless, dull, boring uninteresting cages, into what was to prove to be the rat equivalent of Disneyland. Their new homes had ramps, wheels, slides, swings and other rats. At first the new rats hid under the toys but once they learned that there was nothing to fear, they began to explore their new homes, to use the slides, ramps and swings and to socialize with the other rats. They lost their excess weight, became more active and seemed to enjoy their new surroundings.

When the brains of these rats were examined, it was concluded that nerve cells in the cerebellum each had on average 2000 MORE synapses compared to old rats left in their old cages. The forest of pruned weeping willows had grown new branches. This was fascinating research because it showed that individual nerve cells are capable of growing new filaments even in old age, and this is what happens when the brain is exercised and stimulated. The significance of this fact is obvious for the mind of the elderly dog. Although there is a genetic clock that has programmed the ultimate lifespan of the dog, senescence of the mind is still influenced by the dog's environment. Poor health and lack of mental and physical activity are controllable factors. We can alter the decay of memory by providing our dogs with mental stimulation. In that sense, 'Let sleeping dogs lie' is an unfortunate adage. It is correct in its inference that elderly dogs are more irritable (and elderly dogs do sleep more because of changes in production of the neurotransmitter substance serotonin), but misleading in its suggestion that with age it is best to leave dogs to themselves. This is not so. If we stimulate their minds, we can alter the ageing process and make life more interesting for them in their latter years. The brain is built on instructions from genes but is then modified by events throughout life. Even the ultimate behaviour changes of ageing can be altered by improving the dog's environment.

Illness and the Mind

The adage, 'Let sleeping dogs lie' has a direct application to the mind of the ill dog but here the admonition is a sensible one. In 1987, Professor

Benjamin Hart provided a lucid and sensible explanation for the behaviour of sick animals. All veterinarians in clinical practice know that
the most common reason why a dog owner brings a sick pet to the
veterinary clinic is because of behaviour changes and that the most
commonly described change is, 'dull, lethargic, no appetite'. I hear this
description more than any other. An owner will sometimes describe her
dog as 'depressed' or simply 'not himself'. He will be described as not
wanting to play, not showing his usual activity, looking mournful,
sleeping all day. All the descriptions illustrate subdued behaviour on the
part of the dog, behaviours that when put together often suggest a state
of mind that is akin to depression in humans. We often think that this
state of mind is a RESULT of the illness that the dog has acquired, but it
can be lucidly argued that rather than these being a result of an illness,
that it is an adaptive behavioural stategy on the part of the dog to help
him overcome his illness. The argument goes this way.

Throughout the entire evolution of time, all living creatures have
been constantly and unremittingly exposed to disease causing microbes
such as viruses and bacteria. One consequence of that constant exposure
is that we have evolved a complex immune system to help us either to
prevent or to survive an infection. But there are other disease fighting
methods that animals have evolved and some of these are behavioural.
These behaviour changes which occur almost immediately when a dog
becomes ill are not caused by the illness itself, but are in fact his first line
of defence against disease. The dog's dullness, lethargy and inappetence
all occur before his immune system has a chance to get into action.
These behaviours in the wild ancestors of our canine companions were
life saving in this way.

One of the most common responses to infection is a rise in body
temperature. This is brought on by the release of a substance called
interleukin-1 from certain white blood cells. This substance acts on the
thermostat in the hypothalamus of the brain, effectively raising its set
point. On its own, this is beneficial because many viruses and bacteria
function best at normal body temperature but this interleukin-1 is
probably beneficial in other ways too.

One of the common behaviour changes of the ill dog is his loss of
appetite. Although it has been assumed that this is caused by his fever, it
has been shown in laboratory animals that interleukin-1, rather than
hyperthermia, induces a loss of appetite. Many pet owners will try to
hand feed their ill dogs and are terribly concerned with this behaviour
change but there is evidence that certainly on a short term basis, loss of
appetite can be a life saving behaviour change for two reasons. The first
is that by not eating, the bacteria that have invaded the dog's body are

starved of the minerals they need. More important, if the dog is not hungry, he won't use up precious energy necessary for moving about looking for food. He can stay in one place, conserve energy, reduce heat loss and fight his infection. Experiments with mice have shown that if they are force fed during a bacterial infection, their survival time is actually decreased and mortality is increased. Similarly, if mice are deprived of food for two to three days before an infection, their survival rates are actually increased.

It would seem then that the state of the dog's mind when he is ill, the dullness, listlessness, lethargy and apparent depression, is in fact a genetically programmed adaptive process to help him overcome his illness. It is inborn in the dog's mind and evolved to assist him conserve energy and body heat to help him recover from infection. As all veterinarians know, many canine diseases have a low mortality rate and recovery from them is rapid even without modern medical care. There are others, however, that are far more serious, diseases with a high rate of mortality and these need our active and immediate medical intervention. The dog's problem is that his body can't differentiate between the less serious and more serious illnesses and his response to both is the same. That's why I frequently hear the same description for mild and for serious illnesses – anorexia, lethargy and listlessness.

One conclusion that we can reach from the information that we now have is that the behaviour of the sick dog should be respected. He should be allowed to remain still and should be kept warm. He should not be force fed unless there are specific clinical indications for doing so and his body temperature should be allowed to moderately rise. Drugs to reduce a dog's fever and to return him to his normal behaviour should only be used if his body temperature is becoming excessively high. Body temperature and the behaviour changes that accompany it are our best monitors of the course of infections in dogs. We should treat to eliminate the cause of the infection while respecting the evolutionary behaviour that the dog has developed to help him fight off the infection.

There are many organic and infectious diseases that alter the dog's mind and these are described in depth in veterinary medical textbooks. Infectious diseases of the brain such as rabies or encephalitis have an obvious consequence while metabolic diseases of the endocrine system can cause over or under production of the body's hormones. By altering the harmony of the interrelationships between the glands of the body, these hormonal changes have a direct effect on the dog's behaviour. Too much thyroid hormone, for example, will result in overactivity. Too little will result in lethargy.

Other metabolic diseases can also alter behaviour. Diseases of the liver can have an effect on the dog's mind by causing an inflammation to the brain. Interference in kidney function can result in a build up of body waste products and this can affect the dog's behaviour. Sometimes a behaviour as simple as itchiness and scratching can be traced back to a metabolic disorder in kidney function. Contamination of the dog's body by pollution can dramatically affect his mind. Lead poisoning, for example, can lead to behaviour changes that range from the subtle to the extreme. In any medical examination of the dog's mind, it is always necessary to remember that there are numerous organic reasons for behaviour changes. As well as infections and metabolic diseases and toxins such as lead, tumours of the brain or the endocrine system can alter behaviour patterns. Injuries can do the same, especially brain injuries. There is also a growing list of inherited behavioural disorders in dogs such as flank sucking in Dobermanns or whirling-circling-tail chasing in bull terriers. Different types of epilepsy also appear to be inherited in some breeds. Some of the breeds with dome shaped heads, breeds such as the Chihuahua and the Pekinese have a higher incidence of both typical grand mal epilepsy but also of what is called psychomotor epilepsy, a form of epilepsy in which the dog behaves in a stereotyped but otherwise normal way. I have seen and treated psychomotor epilepsy that has manifested itself in dogs that have chased non existent flies, that have eaten non existent food or that have impulsively chased their tails. If these behaviours are a consequence of true epilepsy, they respond to anticonvulsant drugs such as phenobarb.

One of the most difficult medical conditions to treat is hyperkinesis or simple overactivity. These dogs have boundless energy. They are overly excitable and resist restraint. They have short attention spans and can never sit still. They never seem to get acclimatized to certain situations and generally resist training. These dogs will sometimes respond to drug therapy and this is one final consideration in the mind of the ill dog.

Drugs can have a dramatic effect on the dog's mind and although drug therapy is not a subject for this book, I should mention and briefly classify some of them. For example, it was discovered over twenty years ago that giving tranquillizers to genetically nervous pointers made training easier. On the other hand, giving a drug called scopolamine effectively reduced the dog's motivation and made learning more difficult (probably by exhausting the adreno-pituitary axis and causing emotional exhaustion).

Drugs that act on the dog's mind can be classified into six categories.

1. Tranquillizers are divided into drugs that are antipsychotic and anti-

anxiety. These are useful descriptions in human medicine but in veterinary medicine their division is more simply stated as major and minor tranquillizers. These drugs have a calming effect on the dog's mind and generally reduce spontaneous activity. They can be effective at suppressing aggressive behaviours or at counteracting fear and anxiety. Some have a marked sedative effect.

2. Antidepressants are self descriptive. They are also described as antianxiety drugs and can be useful in treating problems such as separation anxiety in dogs. They are also useful in treating behavioural problems that involve displacement activities such as chewing through walls or eating Persian carpets.

3. Stimulants have few uses in canine behaviour disorders although they are useful in the rare cases of narcolepsy, only reported so far in Dobermanns, Labrador Retrievers, Poodles and Dachshunds, in which the dog simply falls asleep for a short period of time (one to three minutes) during which he is oblivious to his surroundings. Shouting at the dog is just as effective. Stimulants can, perhaps surprisingly, be useful in treating overactive hyperkinetic dogs. They are certainly useful when diagnosing a true case of hyperkinesis.

4. Anticonvulsants have an obvious purpose and act directly on the brain. They are useful and necessary in the treatment of true epilepsy and can be used to diagnose behaviour conditions in which the dog's activities might be learned or might actually be a form of epilepsy, conditions like tail chasing or fly catching. If the behaviour diminishes or disappears when the dog is on a short course of very low dose anticonvulsants, that fact can be used to conclude that the problem is one of epilepsy rather than conditioning.

5. Hormones are perhaps the most frequently used drugs in the field of canine behaviour. There are several, but the most effective and the ones that are most commonly used are synthetic varieties of the female hormone progesterone. (The other female hormone estrogen has little effect on the mind of the male dog and the male hormone testosterone has surprisingly little effect on the mind of the adult female dog.) The progesterone-like drugs are used to suppress undesirable male behaviour such as urine marking, mounting and intermale dominance and for their calming and sedative effects, reducing aggression, anxiety and roaming. They are of little or no benefit in treating territorial or fear aggression.

6. Narcotics are infrequently used in veterinary medicine simply because their use requires the keeping of special medical records and their indications can be met through the use of other drugs such as tranquillizers.

Of course, a final consideration in any investigation of the dog's mind is whether or not he has a concept of death. Until very recently, we have firmly believed that the human is the only sentient animal; that only we are aware of our own identities. This is a cultural phenomenon. Only a few years ago, when I conducted a study of the practising British veterinarian's attitude towards death in general and towards euthanasia in particular, one out of every four practitioners still felt that dogs were not sentient animals. When a similar survey was carried out among Japanese veterinarians, not a single practitioner felt that dogs were not sentient. Our attitudes are defined through our cultural upbringing.

Recent experiments with primates have proved beyond question that chimpanzees and gorillas are sentient animals. The language experiments are classic examples and proved that primates can think in sentences although they are not capable of using grammar.

Washoe was the first chimpanzee to learn American sign language, eventually learning 132 signs. Using this vocabulary, she was able to generalize her signs and invent new ones. Many other chimps followed. Jane Goodall tells the story of Lucy, a chimp raised in a human household at the University of Oklahoma and taught sign language who, after eleven years, was sent to a chimpanzee reserve in the Gambia. Lucy went through a process of rehabilitation to introduce her to a natural environment. When, two years later, someone from her former life visited Lucy, she ran up to the fence of her enclosure and signed, 'Please. Help. Out.' (Through dedicated rehabilitation, Lucy was eventually successfully acclimatized and reintroduced to the wild and died of natural causes several years later.)

David Premack at the University of Pennsylvania used magnetic plastic symbols on a board in his work with the chimp Sarah, confirming that sentience is not simply a human trait. Sarah understood abstracts. She knew the difference between 'same' and 'different'. She understood the difference between 'Sarah give apple to Mary' and 'Mary give apple to Sarah'. Herb Terrace at Columbia University, working with the marvellously appropriately named Nim Chimpsky, came to the same conclusion. Although these experiments set out to investigate whether chimps could learn a language, what they did in fact reveal was that primates are capable of understanding the world around them and of representing that world symbolically in their minds.

These are primates however, of vastly superior intellectual capacity to dogs. The question remains as to whether dogs have such abilities and can by inference have a concept of death. An experiment with pigeons, a species that is intellectually inferior to the dog provides a clue to the answer to this question.

Richard Herrnstein at Harvard University trained pigeons to peck at a panel and get a food reward when he showed them a 35mm slide containing a tree. He did so with a collection of eighty slides, some of which had trees. Herrnstein's pigeons learned to peck for their reward almost immediately. They continued to do so when they were shown slides they had never seen before as long as the new slide contained a tree. And it didn't matter whether the tree was a different species or whether it was in the foreground or background or even in silhouette. The experiment showed that even pigeons are capable of forming categories in their minds.

No similar experiments have yet been carried out on the dog's mind but, by extrapolating from other animal research, it is a safe assumption that dogs have the ability to represent important features of the world about them in symbolic forms in their minds. What they lack is the capacity to express themselves in any descriptive fashion.

Without that ability we will never know whether death is a concept that the dog understands. With our increasing knowledge of the sophistication of animal behaviour, it would be foolish to discount the possibility. What can be stated quite categorically is that dogs do not have an apparent fear of death. That seems to be a purely human phenomenon. I see and hear of few dogs that actually die naturally. Most of my patients either die traumatically, die of illness or disease or are humanely destroyed but of those that do die on their own, I often hear from their owners that the dog sought seclusion at the end by going into another room or crawling under a piece of furniture or wandering into a cupboard. The final act that is played out in the dog's mind is one we might never understand.

Because we have taken the dog into our homes, and our hearts, we have the opportunity to observe his behaviour more closely than we can observe any other animal species. This is a double edged sword for we often make the mistake of interpreting his behaviour in human rather than canine terms. We forget his ancestry. In many ways dog behaviour is remarkably similar to human behaviour and we can probably learn something about ourselves by observing some of the untarnished activities of the dog's mind. The fact remains that the dog's mind works in its own specific ways and with objectives that are not necessarily obvious to us. By understanding these differences, we can become more responsible for our pets' behaviour and at the same time make their lives more stimulating and fulfilling.

Appendix

TEACHING COMMANDS

In his mind, your dog should always look upon you as the alpha dog, the leader of the pack. Remember, don't let your democratic ideals get in the way of proper training. These aren't party tricks you will be teaching. They are the basis for a successful and satisfying relationship with any dog and once trained, he will be as happy and contented to obey your commands as you will be proud of his willingness to do so.

Training can begin as soon as you acquire your pup, as early as seven to eight weeks of age. Get him used to wearing a collar, a flat or rolled leather one is often best. Over the next several weeks you will be teaching your dog to obey several commands, but can use the same simple word to release him from his commands. The simplest release word to teach is 'OK'. And of course, the best negative word is 'NO'.

Dogs should be taught both verbal and visual commands at the same time. Never 'ask' your dog to do something. Don't issue requests with question mark endings. Always use commands that are given fairly and with confidence.

SIT

1. Start training on the first day you have your pup.
2. Let him sniff his food in his bowl. Each time he is fed, hold his bowl above his head in such a way that he is most likely to sit down to keep his eyes on it. Say 'SIT' while he does so.
3. If he doesn't sit on his own, use your hand and gently push his rump down.
4. Reward him for sitting by giving him his meal.

The command to 'SIT' should not always be coupled with a food reward. Carry out the same simple training three or four times daily, where there are no distractions, by offering praise as his reward.

1. Go to a quiet area and hold a toy or simply snap your fingers above your pup's head while at the same time commanding him to 'SIT'.
2. When he does so calmly, reward him with praise. Don't wind him up with excitement. If he is too excited you will lose his attention.
3. Release him from his 'SIT' position by saying 'OK' and always finish off a training session with praise and activity.

4. Once he has learned to 'SIT' in quiet surroundings, move to different areas including from indoors to outdoors, so that he learns to obey command in different situations.

STAY

Once more start training indoors in a quiet place where there are no distractions. You can assure yourself that your pup is more likely to obey if he wears his collar and lead. In that way, you retain control over him. Remember, he won't have the foggiest idea what 'STAY' means when you first use the command. Never discipline him during training for disobeying. Simply go back to the previous stage and repeat the exercise.

1. Crouch down and tell your pup to 'SIT' and when he does so, then tell him to 'STAY'.
2. Use a flat tone of voice and while issuing your command, turn the palm of your hand towards your pup and swing it towards him until it's almost touching his nose.
3. Stand up, keeping the palm of your hand near his nose, then draw back while maintaining your hand's visual impact.
4. After a few seconds release him from his 'STAY' by saying 'OK'.

Needless to say he won't obey at first because he won't know what you're trying to do. Don't get angry with him and don't think he's stupid either. He's just starting to learn. Firmly but patiently say 'NO' when he moves, walk him back, tell him to 'SIT', and repeat the steps I've just outlined.

He might also find it difficult to understand that you simply want him to 'SIT' and might lie down instead. This might please you but if you let him do so, you're teaching him a faulty command. He should only sit when you command him to 'SIT'. If he lies down, do the following.

1. Tell him 'NO' when he lies down and, using, slight tension on his lead this time, repeat the commands to 'SIT' and 'STAY'.
2. If he breaks your command by lying down, be patient and repeat the series once more.
3. If he is obviously flagging, remember to always finish training on a positive note and simply finish off with a command you know he will obey, 'SIT'.

4. Once he will 'SIT' and 'STAY' on command with your standing near him, gradually begin to get farther away from him until he is willing to obey when you are standing at the farthest end of his lead. Always use the lead during training. That way you are assured of having your dog under your ultimate control.
5. As he comes to understand the meaning of the word 'STAY', lengthen the time he will stay but do so erratically. Command him to stay for a minute, for 30 seconds, for two minutes. Be unpredictable in the duration of the command.
6. Once he is well trained in a quiet setting, move to a more stimulating environment such as the garden or the street.
7. If he backslides in his training, return to the previous success level and always finish training on a positive note.

COME

All pups will willingly come to their owners but just because they will as pups doesn't mean they will as adults. Training should start as soon as your new pup arrives in your home, but must be reinforced, especially as dogs mature.

1. Put your pup down, move a few metres away from him then wave your arms and call him by name by saying 'ROVER! COME'. He will almost instinctively do so.
2. Reward him with affection or even a food reward such as a vitamin tablet.
3. Continue to train in this way, using his lead if necessary until he reliably comes to you each time he is called.
4. Once he is well trained in a quiet environment, move into a more stimulating one outdoors and repeat the same series of commands but using his lead so that there is no possibility of disobeying.
5. Once he fluently obeys your command to 'COME' to his name while he is at the end of his lead, graduate to the same command without the safety of his lead.
6. Never train your pup to 'COME' to command or even after training call him in this way if something that he dislikes, like discipline, or a bath is going to ensue.
7. Let me repeat this rule because it is so important. NEVER CALL YOUR DOG TO PUNISH HIM. If punishment is due, go to him rather than have him come to you.

In most instances you will find the command to 'COME' an easy one to teach, until your dog reaches puberty. This is when the pliable pup can become a stubborn teenager. He will then need firm handling if you want him to continue to obey this command. This is in fact when he will really be learning to obey the COMMAND to 'COME' rather than simply following his urge to be with you.

1. Using your lead to assure control, command your dog to 'SIT' and 'STAY'.
2. Move six feet away from him and using his name, command, 'ROVER! COME'.
3. Open your arms to him or in any other way stimulate him to come to you.
4. Praise him with touch or sometimes with a titbit of food.
5. Continue this routine but alter the sequence of commands. Tell him for example, to 'SIT' when he nears you.
6. Repeat this series of commands in all variations so that your dog does not become habituated to a specific series of commands and once this has been done, proceed to issuing commands when he is off his lead.
7. If he does not obey off his lead, revert back to his previous successful level of training, command him to 'SIT' and 'STAY', go to him, put his lead on and saying 'COME' repeatedly, walk him to where you were previously standing. Both of you will find this immensely boring but follow up with a few 'SIT-STAY' commands at that site.
8. Always praise you dog when he comes. Use the same words such as 'Good dog!' or 'Good Rover' and a positive tone to your voice, different to your voice of displeasure.
9. Once more, always finish training on a high note, even if you secretly want to strangle your dog because he's been concentrating on the fluffy young retriever rather than on your commands. If he has been wilfully disobedient to your command to 'COME', finish off training with 'SIT', 'STAY' and a reward from you for his doing so. But keep the reward perfunctory.

DOWN

This is an important command for two reasons. First of all, it teaches your dog to drop on command wherever he is. If he's chasing a squirrel out onto the street, this might save his life. But equally importantly, dropping down is a sign of subservience in dogs and training your dog to obey the command 'DOWN' is an effective way of showing your dominance over him, something that can be exceedingly important if he is a naturally dominant individual. Teaching the command 'DOWN' reinforces your position as leader of the pack and teaches your dog to be calm. Dominant dogs should be commanded to stay 'DOWN' several times a day and for at least fifteen minutes at a time!

1. Command your pup to 'SIT'.
2. Pat the floor and command your pup 'DOWWWWN'. If you pat directly under him, he might lie down simply to sniff your hand.
3. Praise your pup if he obeys your command.
4. If he does not obey, gently draw his forelegs forward while repeating the command, 'DOWN', and then reward him with stroking and verbal praise.
5. Once he is comfortably down, command him to 'STAY' and praise him for doing so.
6. Get up, back away, then say 'OK' to release him from the command.
7. Carry out this procedure only a few times a day to begin with. Don't rush things.
8. Add the 'DOWN' and 'DOWN-STAY' commands to the repertoire of 'SIT', 'STAY' and 'COME' commands that your pup is integrating into his mind.
9. Once this command is obeyed indoors off his lead, move outdoors to a more stimulating environment and repeat the training procedure initially with his lead on.
10. Don't release him from a 'DOWN-STAY' simply because he looks restless. He will quickly learn how effective fidgeting can be.
11. Always reinforce training. Whenever you are out with your dog, mix and match the commands 'SIT', 'STAY', 'COME' and 'DOWN'. Do so randomly so that he must constantly be thinking about what you are telling him to do. In this way he will always look upon you as his leader. He will respect your mind and you will have the postive influence on his that will make him a delightful, responsive and intelligent companion.

If behaviour problems develop here are two basic ways to retrain your dog. The first involves training him to do something else instead of acting in his objectionable way, and this is called counter-conditioning.

COUNTER-CONDITIONING

1. Identify the specific circumstances of the situation in which your dog acts objectionably. For example, if he barks when someone comes to your door, does he do so for everyone or just when certain people come to the door.
2. Decide upon a behaviour that you want your dog to do rather than the one he is doing. For example, if he barks when visitors arrive, you might want him to hold a toy in his mouth instead.

3. Train him to perform the desired behaviour on command and reinforce this new behaviour with a powerful reward (such as smelly cheese). Do this initially every time he carries out the behaviour and then intermittently.
4. Under controlled circumstances, present him with the stimulus that initially caused his unacceptable behaviour (such as someone approaching the door). If you can control the situation, try to keep the evocative stimulus at the lowest possible intensity. Before the dog reacts to the stimulus (i.e. when he pricks up his ears, immediately get his attention and have him carry out his new behaviour. Reward him for obeying. (You might need a whistle at first to get his attention.)
5. Keep training him by this method of counter-conditioning, increasing the intensity of the stimulus until he behaves properly when the evocative stimulus occurs at its normal intensity.

A second method of retraining is called desensitizing. In this method of training, the dog is taught to calmly accept the stimulus that has been previously provoking his unacceptable behaviour.

DESENSITIZATION

1. The objective is to stop a dog from responding to a specific stimulus. For example, if a dog has developed a fear of certain noises, the objective is to make those noises less fearful.
2. Train your dog to lie down on command for at least fifteen minutes. This is best done under highly specific circumstances such as on a specific rug.
3. Initially reward his behaviour with simple food rewards given every fifteen or twenty seconds at first, then gradually at greater intervals.
4. Identify the specific circumstances that elicit the unacceptable response and make sure that you can control these circumstances. For example, if your dog has a fear of loud noises, it is best to use a recording of them in any desensitization.
5. Make sure that your controlled stimulus (e.g. recording) actually does provoke your dog to act in his unacceptable way when it is used at full intensity.
6. Command your dog to lie down and introduce the stimulus at its lowest intensity. Reward your dog for quiet behaviour.
7. Over a period of days and weeks, gradually increase the intensity of the stimulus always rewarding quiescence.
8. If at any time your dog shows objectionable behaviour, only stop the training at a previous level where he behaved acceptably. Always finish a training session on a positive note.

Bibliography

INTRODUCTION

For further reading on wolf behaviour, these books proved useful:
The Carnivores, R. F. Ewer, Cornell University Press, Ithica & London, 1985
Behaviour of Wolves, Dogs and Related Canids, Michael W. Fox, Jonathan Cape, London, 1971
Of Wolves and Men, Barry Holsten Lopez, Charles Scribner's Sons, New York, 1978
The Wolf: the ecology and behaviour of an endangered species, L. David Mech, Natural History Press, New York, 1970

Stephen Jay Gould discusses his theories on neoteny in the last chapter of:
The Mismeasure of Man, W. W. Norton & Co, New York & London, 1979

For added pleasure read his article:
Mickey Mouse Meets Konrad Lorenz, in the Journal Natural History, Vol 88 (5), pp. 30–36, 1979

Other books used were:
The Mind, Anthony Smith, Penguin Books, London, 1985
The Mind Machine, Colin Blakemore, BBC Books, London, 1988

For a fascinating look at dogs from the perspective of a cultural anthropologist, read the chapter 'Perfect Dogs' in:
Belonging in America – Reading between the lines, Constance Perin, University of Wisconsin Press, Madison WN, 1988

CHAPTER ONE: THE GENETICS OF THE MIND

The outstanding book on the genetics of behaviour is the all encompassing:
Sociobiology – The New Synthesis, E. O. Wilson, Harvard University Press, Cambridge, Mass., 1975

Far easier and more exciting, in fact the best defence of Darwinism in print is:
The Blind Watchmaker, Richard Dawkins, W. W. Norton & Co, New York & London, 1987

Still of general interest is the now dated but still very readable:
An Introduction to Animal Behaviour, Aubrey Manning, Edward Arnolds Ltd, London, 1972

For more information on domestic animals read:
The Behaviour of Domestic Animals, Ben Hart, W. H. Freeman & Co, New York, 1985

The most comprehensive and the first scientific look at the genetics of canine behaviour is:
Genetics and the Social Behaviour of the Dog, Scott and Fuller, University of Chicago Press, Chicago, 1965

CHAPTER TWO: THE BRAIN

The most comprehensive text on the anatomy of the dogs' brain is:
Physiological and Clinical Anatomy of the Domestic Animals, A. S. King, Oxford Scientific Publications, Oxford University Press, Oxford, 1987

Further information is available in:
Veterinary Neurology, J. E. Oliver, B. F. Hoerlein, I. G. Mayhew, W. B. Saunders, Philadelphia, London, Toronto, 1987
and Professor Ursin's chapter in:
Nutrition and Behaviour in Dogs and Cats. Proceedings of the First Nordic Symposium on Small Animal Veterinary Medicine, Oslo, September 15–18, 1982. R. S. Anderson (ed), Pergamon Press, Oxford, 1983

CHAPTER THREE: THE SENSES

Most information on canine senses comes from recent scientific journals. There is, however, useful material in several books although most of these are of a specialist nature. I have used:
Animal Behaviour for Veterinarians and Animal Scientists, K. A. Houpt and T. R. Wolski, Iowa State University Press, Ames, Iowa, 1982
The Beagle as an Experimental Animal, R. L. Werner and L. Z. MacFarland, Iowa State University Press, Ames, Iowa, 1970
Current Topics in Developmental Biology, Neurobiology and Neural Development. Vol 21, Academic Press, New York, 1987
Rhythmic Aspects of Behavior, F. M. Brown and R. C. Graeber, Erlbaum Press, Hillsdale, New Jersey, 1982

Professor Hubert Montagner's work on canine scent has been related to his interest in attachment between mothers and infants, work that he explains in:
L'Attachement, Les Debuts de la Tendresse, Editions Odile Jacob, Paris, 1988

Oliver Sacks's peculiar story about the man who smelled like a dog is recounted in:
The Man Who Mistook His Wife For a Hat, Picador Books, London, 1986

There is comparative information on animal senses in:
Animal Behaviour: Selected Topics in Biology. A. P. Blookfield. Thomas Nelson & Sons Ltd, Walton-on-Thames, Surrey, 1980

CHAPTER FOUR: HORMONES AND THE MIND

Two books are good sources on the hormonal influence on behaviour:
Domestic Animal Behaviour for Veterinarians and Animal Scientists, Katherine A. Houpt
& Thomas R. Wolski, The Iowa State University Press, Ames, 1982

and

The Behaviour of Domestic Animals, Benjamin L. Hart, W. H. Freeman & Co, New
York, 1985

For further reading see:
The Beagle as an Experimental Animal, R. L. Weiner & L. Z. MacFarland, Iowa State
University Press, Ames, 1970
The journal *Animal Behaviour*, Balliere Tindall, Academic Press Ltd, Sidcup, Kent
DA14 5HP frequently has articles on sensory and hormonal influences on
canine behaviour

CHAPTER FIVE: COMMUNICATION

Michael Fox did his initial studies on canine behaviour with wolves. Later he worked
with Scott & Fuller at Bar Habour and has gone on to write more prolifically than
anyone else on dog behaviour. Scott & Fuller's text still remains the bible in
understanding the dog's mind. It is called:
Genetics and the Social Behavior of the Dog, University of Chicago Press, Chicago, 1965

Michael Fox's most relevant books concerning communication in dogs are:
The Dog – Its Domestication and Behaviour, Garland STPM Press, New York, 1983
Understanding your Dog, Coward, McCann & Geoghegan Inc, New York, 1982

and the chapter on canine communication in:
How Animals Communicate, M. W. Fox & J. A. Cohen, Indiana University Press,
London, 1977

Konrad Lorenz is a masterly source of information on animal communication. Three
of Lorenz's books still make fascinating reading. They are:
Studies in Animal and Human Behaviour
King Solomon's Ring
Evolution and Modification of Behaviour
All are published by Methuen, London

Play activity is amply covered in:
Animal Play Behaviour, Robert Fagan, Oxford University Press, New York, Oxford,
1981

CHAPTER SIX: EARLY LEARNING – MATERNAL AND PEER IMPRINTING

Early learning in dogs is covered in the previously mentioned *Genetics and the Social
Behaviour of the Dog* by Scott & Fuller and in *The Beagle as an Experimental
Animal*.

One of the most prolific popular writers on the dog's mind is Anders Hallgren but his books are only available in Swedish. For those with an understanding of that language there are three books:

Problemhund och Hundproblem (*Problem Dogs and Dog Problems*), ICA forlaget, Vasteras, Sweden, 1985

Lyckliga Lydiga Hundar (*Happy Obedient Dogs*), ICA forlaget, Vasteras, Sweden, 1984

Hundens Gyllene Regler (*The Golden Rules of the Dog*), Jycke-Tryke forlag AB, Koping, Sweden, 1984

Perhaps easier to read are:

Behaviour Problems in Dogs, William Campbell, American Veterinary Publications, Santa Barbara, California, 1975

The Veterinary Clinics of North America, Small Animal Practice – Animal Behavior, Victoria Voith and Peter Borchelt, Vol 12, No 4, WB Saunders, Philadelphia, Toronto, London, 1982

My previous book on canine behaviour:

Games Pets Play: How not to be manipulated by your pet, Michael Joseph, London and Viking Press, New York, 1986, has practical information on early learning and its influence on the developing mind of the dog.

CHAPTER SEVEN: LATER LEARNING – OUR INFLUENCE ON THE DEVELOPING MIND

Our influence on the development of the dog's mind has been covered in a number of papers presented at scientific meetings. These papers appear in several volumes including:

Interrelations Between People and Pets, Bruce Fogle (ed), Charles C. Thomas, Springfield, Illinois, 1980

New Perspectives on our Lives with Companion Animals, Aaron Katcher & Alan Beck (eds), University of Pennsylvania Press, Philadelphia, 1983

The Pet Connection – Its influence on our health and quality of life, R. K. Anderson, B. & L. Hart (eds), University of Minnesota Press, Minneapolis, 1984

The subject is also covered in my previous book:

Pets and their People, Collins, London 1983 and Viking, New York, 1983

And for readers of Danish, Roger Abrantes has covered this field in:

Hunden vor ven – Psykologi fremfor magt, Borgen, Denmark, 1986

CHAPTER EIGHT: SOCIAL BEHAVIOUR – AGGRESSION

Victoria Voith and Peter Borchelt discuss aggression in the previously mentioned *Veterinary Clinics of North America*.

There is good coverage of the subject in:

The Behaviour of Domestic Animals, Benjamin Hart, W. H. Freeman & Co, New York, 1985

Konrad Lorenz initially investigated the behaviour in:

On Aggression, Methuen, London, 1966

CHAPTER NINE: SOCIAL BEHAVIOUR – EATING, EXPLORING, ELIMINATING

These areas are primarily covered in articles in journals on animal behaviour. Two worthwhile books are:

Domestic Animal Behaviour for Veterinarians and Animal Scientists, Katherine A. Houpt & Thomas R. Wolski, The Iowa State University Press, Ames, 1982

Nutrition and Behaviour in Dogs and Cats, Ron Anderson (ed), Pergamon Press, Oxford and New York, 1982

CHAPTER TEN: SOCIAL BEHAVIOUR – FEARS, PHOBIAS, EXCITEMENT

This topic is discussed in most of the previously cited books on canine behaviour but most lucidly in:

Manual of Canine Behaviour, Valerie O'Farrell. British Small Animal Veterinary Association, Cheltenham, Glos. 1986

CHAPTER ELEVEN: SOCIAL BEHAVIOUR – PACK, SEX, MATERNAL

Canine pack behaviour is discussed in the previously cited wolf literature although it does not necessarily apply to domesticated dogs.

Sex and maternal behaviour is described by Michael Fox and principally by Benjamin Hart. Both of his books are excellent references:

The Behaviour of Domestic Animals, W. H. Freeman & Co, New York, 1985

Canine and Feline Behaviour Therapy, Febiger, Philadelphia, 1985

CHAPTER TWELVE: BREED DIFFERENCES IN BEHAVIOUR

Scott & Fuller's *Genetics and the Social Behaviour of the Dog* was the first text to examine this subject. More recently Benjamin and Lynette Hart have published their research findings in popular form as:

The Perfect Puppy, W. H. Freeman & Co, New York & Oxford, 1988

CHAPTER THIRTEEN: THE MIND OF THE ELDERLY AND THE ILL

Once more, most information on this subject is found in journals on animal behaviour. There are however, several books that are of interest.

Animals, Aging and the Aged, Leo Bustad, University of Minnesota Press, Minneapolis, Minnesota, 1980

Active Years for your Aging Dog, Bernard Hershhorn, Hawthorn Books, New York, 1978

The Biology of Aging, J. A. Behnke, C. E. Finch, G. B. Moments (eds), Plenum Press, New York, 1979

THE CAT'S MIND
by Bruce Fogle

'THE DR SPOCK OF THE CAT WORLD'

We are, says Bruce Fogle, becoming an increasingly cat-loving nation – cats are not as intrusive as dogs, don't upset routines, are clean and relatively independent and satisfy our need to care for living things. However, the main reason why cats are so popular as pets is our interpretation (which is frequently wrong) of their behaviour.

Drawing on both the vast amount of worldwide research into cat behaviour and twenty-five years of observation as a practising veterinarian, Bruce Fogle examines every aspect of feline behaviour from kittenhood to old age and provides a fascinating insight into what really goes on in the cat's mind.

This is a book which will enable everyone who keeps cats as pets to enjoy an even richer relationship with these most attractive yet enigmatic creatures.